New Generation Mathematics Textbook for Grade 2

THE NEW GENERATION MATHEMATICS SERIES

A complete mathematics course for every grade, built on United States standards.

Each grade has a full-color **Textbook** and a matching black-and-white **Workbook**.

New Generation Mathematics Textbook for Grade 2

Developing Fluency, Reasoning, and Real-World Mathematical Thinking

Emma Brightwell

Aligned to United States Common Core State Standards for Mathematics

New Generation Mathematics Textbook for Grade 2
Developing Fluency, Reasoning, and Real-World Mathematical Thinking

This book is aligned to the Common Core State Standards for Mathematics, Grade 2. The Common Core State Standards are a work of the National Governors Association Center for Best Practices and the Council of Chief State School Officers. This publication is an independent work and is not endorsed by, or affiliated with, those organizations.

First edition.
Interior design and illustration by the author.
Printed in the United States of America.

Contents

Contents *continued*

Contents *continued*

A Letter to Families and Teachers

Dear Families and Teachers,

Second grade is the year the numbers get bigger and the thinking gets deeper. Children arrive able to add and subtract within twenty. They leave able to work confidently to a thousand, to measure with real tools, to handle money and time, and — most importantly — to explain *why* a method works.

This textbook follows the United States Common Core State Standards for Grade 2 in full. Every Grade 2 standard is taught explicitly, and the standard code appears at the top of each lesson so you always know exactly what is being learned.

Each lesson keeps the same predictable rhythm as the Grade 1 book. **Let's Learn** introduces the idea with a picture. **Worked Example** shows the thinking step by step. **Try It Together** is guided practice. **On Your Own** is independent practice. **Math Talk** asks the child to explain their reasoning out loud — please do not skip it. In second grade, explaining is itself a standard.

Two habits matter more than speed this year. The first is estimating before calculating. The second is checking: every subtraction can be checked by adding. Both are taught here, and both are worth insisting on.

Helpful materials are simple: a ruler marked in inches and centimeters, a handful of real coins, and a clock with hands. Complete answers begin on the Answer Key page listed in the contents.

A Letter to You

Hello again, mathematician!

Last year you learned to count, add, and subtract. This year the numbers grow. You will work with hundreds. You will handle money, read a clock to the minute, and measure real things with a real ruler.

Here is the most important thing in this whole book: getting the right answer is only half the job. The other half is being able to say *how you know*. When you reach a **Math Talk** box, say your thinking out loud to someone. If nobody is nearby, say it to yourself.

You will make mistakes. Every mathematician does. A mistake is not a failure — it is a signal showing you exactly which step to look at again. Find it, fix it, and carry on.

Take your time. Use your fingers, blocks, or drawings for as long as they help. There is no prize for rushing.

Let's begin.

How to Use This Book

Every unit in this book has the same shape, so you always know where you are.

THE SHAPE OF A UNIT

Unit opener — the learning goals and the standards the unit covers.
Five lessons — each one fills two pages.
Math in My World — one real situation with several questions.
Unit Review — two pages that check everything, ending with a challenge.

THE PARTS OF A LESSON

New Words gives the vocabulary before you need it.
Let's Learn shows the idea with a picture.
Worked Example walks through the thinking one step at a time.
Try It Together is practice to do with someone.
On Your Own is practice to do alone.
Math Talk asks you to explain.
Remember holds a tip worth keeping.

Pictures That Do the Thinking

These pictures appear again and again. Each one is a tool for thinking, not decoration.

BLOCKS

Base-ten blocks — a flat is 100, a rod is 10, a cube is 1.

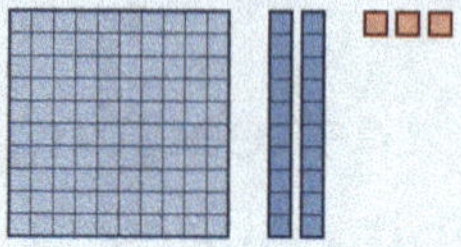

CHART

Place-value chart — shows what each digit is worth.

Hundreds	Tens	Ones
2	4	6

NUMBER LINE

Open number line — jump by tens, then by ones.

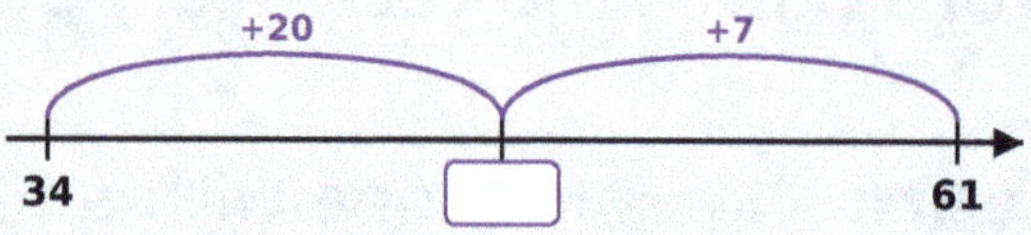

COLUMNS

Column method — the dotted boxes are for regrouping.

Grade 2 Standards at a Glance

These are the United States Common Core State Standards for Grade 2 mathematics. Every one is taught in this book.

OPERATIONS AND ALGEBRAIC THINKING

2.OA.A.1 One- and two-step word problems within 100.
2.OA.B.2 Fluently add and subtract within 20; know all one-digit sums from memory.
2.OA.C.3 Odd and even numbers.
2.OA.C.4 Rectangular arrays and repeated addition.

NUMBER AND OPERATIONS IN BASE TEN

2.NBT.A.1 Hundreds, tens, and ones.
2.NBT.A.2 Count within 1,000; skip-count by 5s, 10s, and 100s.
2.NBT.A.3 Numerals, number names, and expanded form.
2.NBT.A.4 Compare three-digit numbers.
2.NBT.B.5 Fluently add and subtract within 100.
2.NBT.B.6 Add up to four two-digit numbers.
2.NBT.B.7 Add and subtract within 1,000.
2.NBT.B.8 Mentally add or subtract 10 or 100.
2.NBT.B.9 Explain why strategies work.

MEASUREMENT AND DATA

2.MD.A.1 Measure with rulers and other tools.
2.MD.A.2 Measure the same object in two different units.
2.MD.A.3 Estimate lengths.
2.MD.A.4 Find how much longer one object is.
2.MD.B.5 Word problems involving length.
2.MD.B.6 Number line diagrams.
2.MD.C.7 Tell time to five minutes, with a.m. and p.m.
2.MD.C.8 Money: dollars and cents.
2.MD.D.9 Line plots of measurement data.
2.MD.D.10 Picture graphs and bar graphs.

GEOMETRY

2.G.A.1 Recognize and draw shapes by their attributes.
2.G.A.2 Partition a rectangle into rows and columns of squares.
2.G.A.3 Halves, thirds, and fourths.

WHY MATH TALK MATTERS

A note on Grade 2: the standards ask children not only to calculate but to *explain*. Standard 2.NBT.B.9 is met in every Math Talk box in this book.

UNIT

1

Place Value to 100

YOUR LEARNING GOALS

In this unit you will learn to...

- Write a two-digit number as tens and ones.
- Count on and back by ones, fives, and tens.
- Compare two-digit numbers with >, =, and <.
- Decide whether a number is even or odd.
- Use rows and columns to find how many in all.

STANDARDS COVERED

2.NBT.A.1 Understand that the digits of a three-digit number represent hundreds, tens, and ones.

2.NBT.A.2 Count within 1000; skip-count by 5s, 10s, and 100s.

2.NBT.A.4 Compare two three-digit numbers using >, =, and <.

2.OA.C.3 Decide whether a group of objects is odd or even.

2.OA.C.4 Use addition to find the total in a rectangular array.

1 Tens and Ones

Standard: 2.NBT.A.1

NEW WORDS

digit — one of the symbols 0 to 9
place value — what a digit is worth because of where it sits

LET'S LEARN

In a two-digit number, the left digit counts **tens** and the right digit counts **ones**.

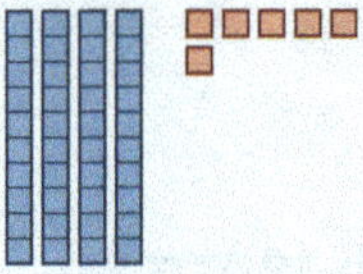

Tens	Ones
4	**6**

4 tens and 6 ones is **46**.

WORKED EXAMPLE

Step 1. Count the rods: 7 tens is 70.

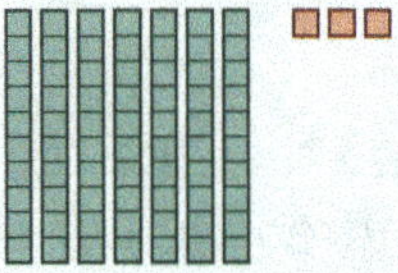

Step 2. Count the cubes: 3 ones is 3.
Answer: 70 + 3 = **73**.

MATH TALK

In the number 58, what is the 5 worth? What is the 8 worth?

TRY IT TOGETHER

Write the number shown.

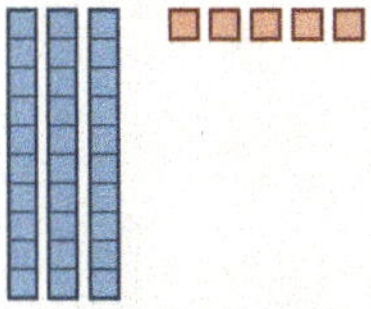

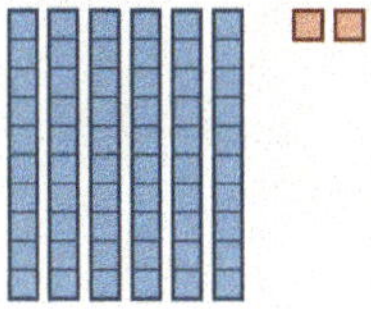

ON YOUR OWN

Write each number as tens and ones.

1 **27 = ☐ tens ☐ ones**

2 **84 = ☐ tens ☐ ones**

3 **60 = ☐ tens ☐ ones**

Write the number.

4 **5 tens 9 ones = ☐**

5 **9 tens 0 ones = ☐**

6 **1 ten 7 ones = ☐**

7 **7 tens 4 ones = ☐**

REMEMBER

The same digit is worth different amounts in different places. In 33, the left 3 is worth 30 and the right 3 is worth 3.

2 Counting Patterns

Standard: 2.NBT.A.2

NEW WORDS

skip-count — to count in equal jumps, such as 5, 10, 15

LET'S LEARN

Skip-counting is faster than counting by ones. Look for the pattern in the digits.

1	2	3	4	**5**	6	7	8	9	**10**
11	12	13	14	**15**	16	17	18	19	**20**
21	22	23	24	**25**	26	27	28	29	**30**
31	32	33	34	**35**	36	37	38	39	**40**
41	42	43	44	**45**	46	47	48	49	**50**
51	52	53	54	**55**	56	57	58	59	**60**

Counting by fives, every number ends in 5 or 0.

WORKED EXAMPLE

Step 1. Start at 30 and count on by tens.

30	**40**	**50**		

Step 2. Only the tens digit changes.

Answer: 60 and 70.

MATH TALK

When you count by tens from 24, what stays the same every time?

TRY IT TOGETHER

Fill in the missing numbers.

15	20		30	
	47	48		50

ON YOUR OWN

Count by fives.

25			40	

Count by tens.

18			48	

Count back by ones.

73			70	

1 **10 more than 56 is** ☐

2 **10 less than 91 is** ☐

3 Comparing Numbers

Standard: 2.NBT.A.4

NEW WORDS

greater than (>) — more than
less than (<) — smaller than

LET'S LEARN

Compare the **tens** first. If the tens are the same, compare the ones.

63 > 58

6 tens beats 5 tens, so 63 > 58.

The open end of the symbol always faces the greater number.

WORKED EXAMPLE

Step 1. Compare 47 and 42. Both have 4 tens.

47 > 42

Step 2. Compare the ones: 7 is more than 2.

Answer: 47 > 42.

MATH TALK

Can a number with fewer tens ever be greater? Explain your thinking.

TRY IT TOGETHER

Write >, =, or < in the box.

1. 38 ☐ 83
2. 56 ☐ 56
3. 70 ☐ 69
4. 24 ☐ 29

ON YOUR OWN

Write >, =, or <.

1. 91 ☐ 19
2. 45 ☐ 54
3. 80 ☐ 80
4. 67 ☐ 62
5. 33 ☐ 38
6. 50 ☐ 49

Order from least to greatest: 72, 27, 70

☐ ☐ ☐

REMEMBER

Reading the symbol out loud helps: 63 > 58 is read “sixty-three is greater than fifty-eight”.

4 Even and Odd

Standard: 2.OA.C.3

NEW WORDS

even — can be split into two equal groups
odd — has one left over when split into two groups

LET'S LEARN

Pair the objects. If nothing is left over, the number is **even**.

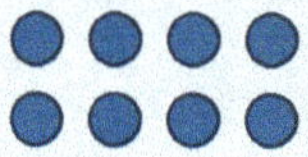

8 makes 4 pairs with none left over. 8 is **even**.

7 makes 3 pairs with 1 left over. 7 is **odd**.

WORKED EXAMPLE

Step 1. Is 14 even or odd? Split it in half.

$$7 + 7 = 14$$

Step 2. 14 is two equal groups of 7.
Answer: 14 is even.

MATH TALK

Every even number can be written as a double. Why does that work?

TRY IT TOGETHER

Write **even** or **odd**.

1 **10** 2 **15** 3 **6** 4 **9**

ON YOUR OWN

Circle every **even** number.

1 **12** 2 **17** 3 **20** 4 **35**

5 **48** 6 **51** 7 **64** 8 **79**

Write each even number as a double.

9 **16 =** ☐ + ☐

10 **18 =** ☐ + ☐

REMEMBER

Look at the ones digit. If it is 0, 2, 4, 6, or 8, the number is even.

5 Rows and Columns

Standard: 2.OA.C.4

NEW WORDS

array — objects arranged in equal rows and columns

repeated addition — adding the same number again and again

LET'S LEARN

An **array** has equal rows. Add the rows to find the total.

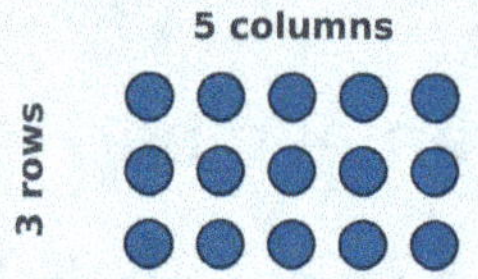

5 + 5 + 5 = 15

3 rows of 5 is 15.

WORKED EXAMPLE

Step 1. Count the rows: 4. Count each row: 2.

Step 2. Add: 2 + 2 + 2 + 2.

Answer: 8 stars.

MATH TALK

Would 5 rows of 3 give the same total as 3 rows of 5? Explain.

TRY IT TOGETHER

Write the addition sentence and the total.

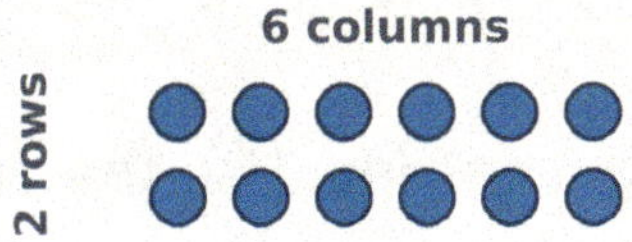

$\square + \square = \square$

ON YOUR OWN

Write the total for each array.

 $\square$ 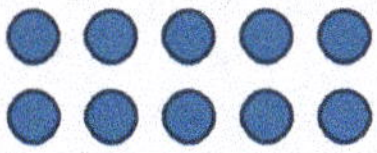 $\square$

Draw an array with 4 rows of 3. Write the total.

Math in My World: The School Garden

Problem solving with Unit 1 skills

Class 2B is planting a garden. Use the pictures to answer.

THE PICTURES

The seed trays

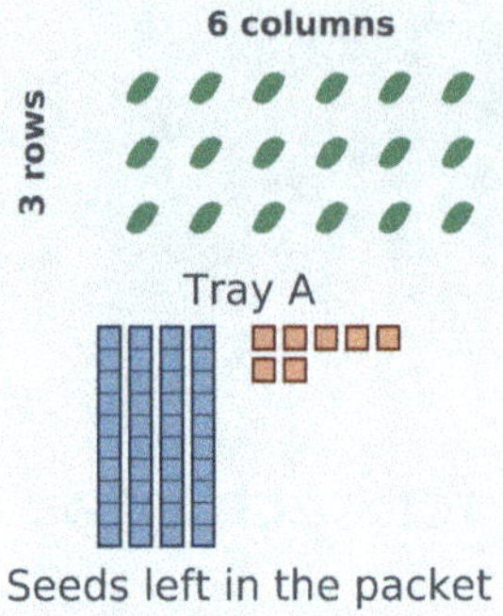

YOUR TURN

1. How many plants are in Tray A? ________
2. Write an addition sentence for Tray A.
3. How many seeds are left in the packet? _____
4. Write that number as tens and ones. _____ tens _____ ones
5. Is the number of plants in Tray A even or odd? ________
6. Tray B has 45 seeds. Which tray has more? ________

MATH TALK

Explain to a partner how an array helped you count the plants faster than counting one by one.

Unit 1 Review

Review of Lessons 1.1–1.5

A Part A · Place value

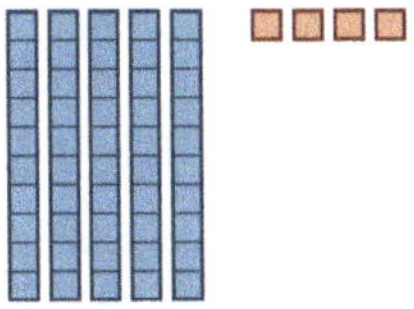

☐

1 **68 = ☐ tens ☐ ones**

2 **3 tens 9 ones = ☐**

B Part B · Counting patterns

35	☐	**45**	☐	**55**

☐	**26**	**36**	☐	**56**

C Part C · Compare

3 **74 ☐ 47**

4 **52 ☐ 58**

D Part D · Even and odd

Write **even** or **odd**.

5 **14** 6 **23** 7 **40** 8 **57**

E Part E · Arrays

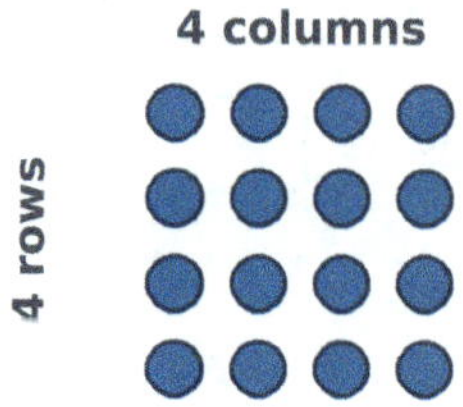

$\square + \square + \square + \square = \square$

CHALLENGE

Use the digits 6 and 3 once each. Write the greatest and the least two-digit number you can make. Then compare them with a symbol.

UNIT

2

Fact Fluency within 20

YOUR LEARNING GOALS

In this unit you will learn to...

- Use doubles and near doubles quickly.
- Make a ten to add across 10.
- Use addition facts to subtract.
- Write the four facts in a fact family.
- Add three numbers by looking for easy pairs.

STANDARDS COVERED

2.OA.B.2 Fluently add and subtract within 20 using mental strategies; know from memory all sums of two one-digit numbers.
2.OA.A.1 Use addition and subtraction within 100 to solve word problems.

1 Doubles and Near Doubles

Standard: 2.OA.B.2

NEW WORDS

double — adding a number to itself

near double — a fact that is one away from a double

LET'S LEARN

Doubles are worth memorizing. They unlock the facts next to them.

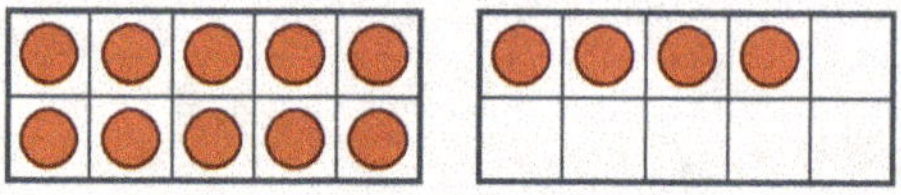

$$7 + 7 = 14$$

Then 7 + 8 is just one more: 15.

WORKED EXAMPLE

Step 1. Find 6 + 7. The nearest double is 6 + 6 = 12.

$$6 + 6 = 12$$

Step 2. 7 is one more than 6, so add one more.

Answer: 6 + 7 = 13.

MATH TALK

Which double would you use to solve 8 + 9? Why?

TRY IT TOGETHER

Use the double to help you.

1 **5 + 5 = ☐ so 5 + 6 = ☐**

2 **9 + 9 = ☐ so 9 + 8 = ☐**

ON YOUR OWN

Add.

1 **4 + 4 = ☐**

2 **6 + 6 = ☐**

3 **8 + 8 = ☐**

4 **7 + 8 = ☐**

5 **5 + 6 = ☐**

6 **9 + 10 = ☐**

Write the double you would use.

7 **for 4 + 5, use ☐ + ☐**

REMEMBER

Doubles to know by heart: 1+1, 2+2, 3+3, 4+4, 5+5, 6+6, 7+7, 8+8, 9+9, 10+10.

2 Make a Ten

Standard: 2.OA.B.2

NEW WORDS

make a ten — break a number so the first number reaches 10

LET'S LEARN

Ten is easy to add to. Break the second number so the first one fills a ten.

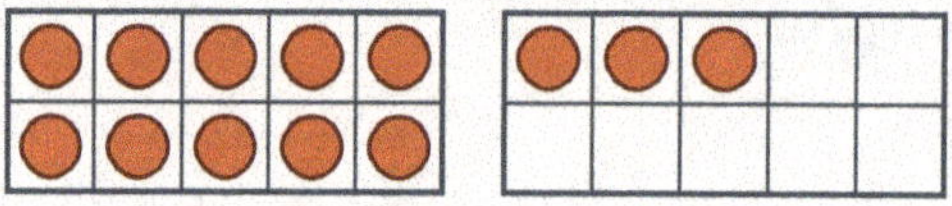

$$8 + 5 = 8 + 2 + 3$$

$$10 + 3 = 13$$

WORKED EXAMPLE

Step 1. Find 9 + 6. 9 needs 1 more to make 10.

$$9 + 1 = 10$$

Step 2. 6 is 1 and 5, so 5 is left.

Answer: 10 + 5 = 15.

MATH TALK

Why is it easier to add 10 + 5 than 9 + 6, even though the answer is the same?

TRY IT TOGETHER

Break the number to make a ten.

1 $7 + 5 = 7 + 3 + \square = \square$

2 $8 + 6 = 8 + 2 + \square = \square$

ON YOUR OWN

Add.

1 $9 + 4 = \square$

2 $8 + 7 = \square$

3 $7 + 6 = \square$

4 $9 + 8 = \square$

5 $6 + 8 = \square$

6 $9 + 9 = \square$

Show 8 + 9 on the open number line.

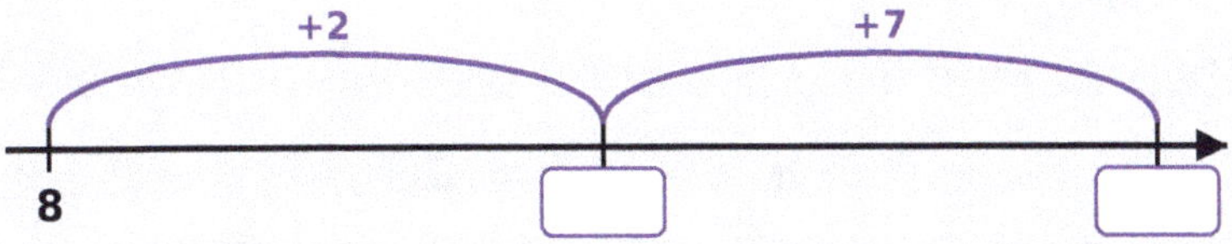

3 Subtraction Facts

Standard: 2.OA.B.2

NEW WORDS

difference — the answer when you subtract

LET'S LEARN

To subtract, think of the addition fact that goes with it.

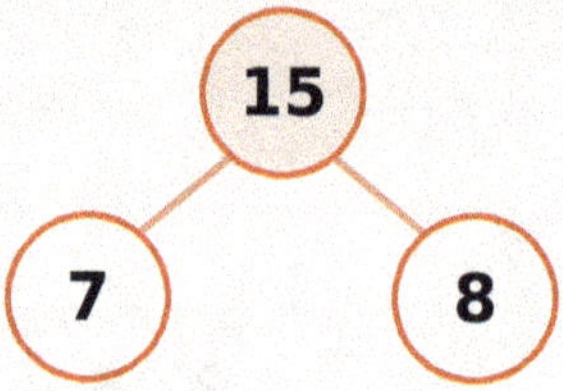

$$7 + 8 = 15$$

$$15 - 7 = 8$$

The bond holds all four facts.

WORKED EXAMPLE

Step 1. Find 13 – 5. Ask: 5 plus what makes 13?

$$5 + \square = 13$$

Step 2. 5 + 8 = 13.

Answer: 13 – 5 = 8.

MATH TALK

How does knowing 6 + 7 = 13 help you find 13 – 6?

TRY IT TOGETHER

Use the addition fact to subtract.

1 **8 + ☐ = 14 so 14 − 8 = ☐**

2 **9 + ☐ = 16 so 16 − 9 = ☐**

ON YOUR OWN

Subtract.

1 **12 − 4 = ☐**

2 **17 − 9 = ☐**

3 **15 − 6 = ☐**

4 **11 − 3 = ☐**

5 **18 − 9 = ☐**

6 **14 − 7 = ☐**

Subtract ten first, then the extra ones.

7 **16 − 10 = ☐**

8 **19 − 10 = ☐**

REMEMBER

Every subtraction fact has a partner addition fact. Finding the partner is often the fastest route.

4 Fact Families

Standard: 2.OA.B.2

NEW WORDS

fact family — the four facts made by three related numbers

LET'S LEARN

Three numbers make two addition facts and two subtraction facts.

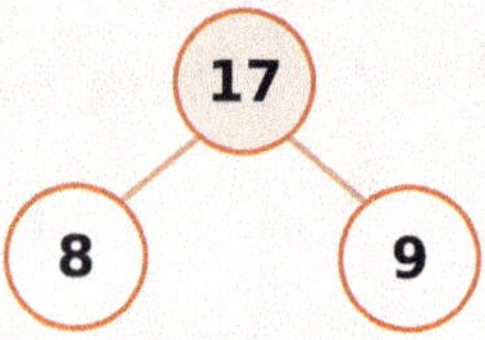

8 + 9 = 17

9 + 8 = 17

17 − 8 = 9

17 − 9 = 8

WORKED EXAMPLE

Step 1. Use 5, 7, and 12.

Step 2. The whole is 12. The parts are 5 and 7.

Answer: 5 + 7 = 12, 7 + 5 = 12, 12 − 5 = 7, 12 − 7 = 5.

MATH TALK

Why do the two subtraction facts in a family always start with the same number?

TRY IT TOGETHER

Write the four facts for 6, 8, and 14.

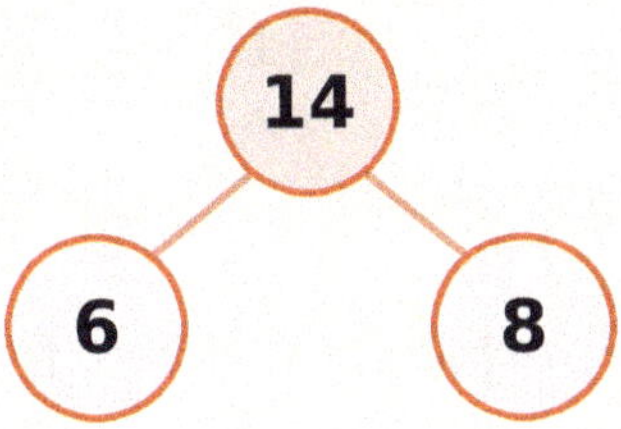

ON YOUR OWN

Fill in the missing number.

1 $7 + \square = 15$

2 $\square + 9 = 18$

3 $13 - \square = 5$

4 $\square - 6 = 8$

Write the missing member of each family.

5 4, 9, ☐

6 7, ☐, 16

5 Adding Three Numbers

Standard: 2.OA.A.1

NEW WORDS

group — to join two numbers first because they are easy

LET'S LEARN

You may add in any order. Look for a pair that makes 10 or a double.

6 + 4 + 9

6 + 4 makes 10, so the sum is 10 + 9 = **19**.

WORKED EXAMPLE

Step 1. Find 3 + 8 + 7. Look for a ten.

3 + 7 = 10

Step 2. Add the 8 that is left.
Answer: 10 + 8 = 18.

MATH TALK

Does changing the order of the three numbers change the total? Why or why not?

TRY IT TOGETHER

Circle the pair you add first, then find the total.

1 $5 + 5 + 6 =$ ☐

2 $2 + 9 + 8 =$ ☐

ON YOUR OWN

Add.

1 $4 + 6 + 7 =$ ☐

2 $3 + 3 + 9 =$ ☐

3 $8 + 2 + 5 =$ ☐

4 $6 + 6 + 4 =$ ☐

5. A shelf holds 7 red, 3 blue, and 8 green books. How many books in all? ______

REMEMBER

Adding is friendly: you can rearrange the numbers freely and the total stays the same.

Math in My World: The Sports Day

Problem solving with Unit 2 skills

Class 2B is keeping score at sports day.

THE PICTURE

Points scored

Red team	★★★★★★★★
Blue team	★★★★★★
Green team	★★★★★★★★★

Key: each symbol = 1

YOUR TURN

1. How many points did the red team score? ______

2. How many points did red and blue score together?

☐ + ☐ = ☐

3. How many more points did green score than blue? ______

4. How many points did all three teams score in all? ______

5. Green scored 5 more points after lunch. What is green's new total? ______

MATH TALK

For question 4 you added three numbers. Which two did you add first, and why did you choose those?

✓ Unit 2 Review

Review of Lessons 2.1–2.5

A Part A · Add within 20

1 $7 + 7 = \square$

2 $8 + 9 = \square$

3 $6 + 5 = \square$

4 $9 + 7 = \square$

B Part B · Subtract within 20

5 $14 - 6 = \square$

6 $16 - 8 = \square$

7 $12 - 5 = \square$

8 $18 - 9 = \square$

C Part C · Fact family

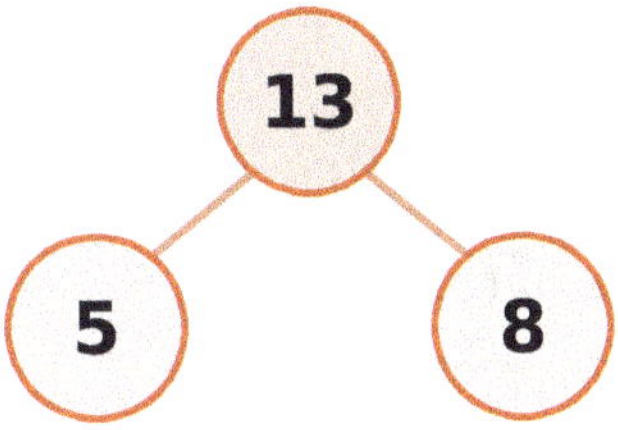

D Part D · Three numbers

9 $5 + 5 + 8 = \square$

10 $3 + 7 + 6 = \square$

E Part E · Missing numbers

11 $9 + \square = 17$

12 $\square - 7 = 6$

F Part F · Word problem

13. Sam had 15 marbles. He gave 7 to a friend and then found 4 more. How many marbles does he have now?

CHALLENGE

Find two different pairs of one-digit numbers that add to 13. Then write a subtraction fact for each pair.

UNIT

3

Adding Two-Digit Numbers

YOUR LEARNING GOALS

In this unit you will learn to...

- Add tens to tens and ones to ones.
- Add on an open number line.
- Break numbers apart to add them.
- Regroup ten ones into a new ten.
- Add three or four two-digit numbers.

STANDARDS COVERED

2.NBT.B.5 Fluently add and subtract within 100 using strategies based on place value and the properties of operations.

2.NBT.B.6 Add up to four two-digit numbers using strategies based on place value.

2.NBT.B.9 Explain why addition and subtraction strategies work.

1 Adding Tens and Ones

Standard: 2.NBT.B.5

NEW WORDS

sum — the answer when you add

LET'S LEARN

Add the tens together, then add the ones together.

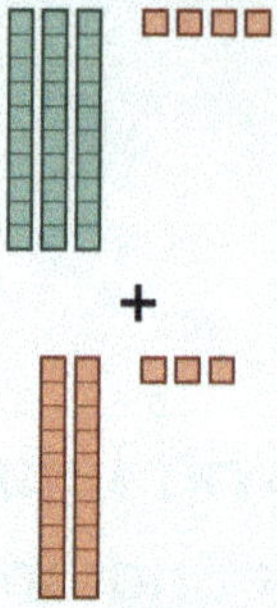

$$30 + 20 = 50$$

$$4 + 3 = 7$$

$$50 + 7 = 57$$

WORKED EXAMPLE

Step 1. Find 42 + 35. Add the tens: 40 + 30 = 70.
Step 2. Add the ones: 2 + 5 = 7.

$$70 + 7 = 77$$

Answer: 42 + 35 = 77.

MATH TALK

Why is it safe to add the tens before the ones?

TRY IT TOGETHER

Add the tens, then the ones.

1 **23 + 45 = ☐** 2 **51 + 36 = ☐**

ON YOUR OWN

Add.

1 **14 + 22 = ☐** 2 **35 + 41 = ☐**

3 **60 + 27 = ☐** 4 **13 + 13 = ☐**

5 **52 + 34 = ☐** 6 **46 + 23 = ☐**

7. Write 25 + 62 as tens plus ones, then find the sum.

20 + 60 + **5 + 2**

☐

2 The Open Number Line

Standard: 2.NBT.B.5

NEW WORDS

open number line — a number line with only the numbers you need marked on it

LET'S LEARN

Start at the first number. Jump by tens, then by ones.

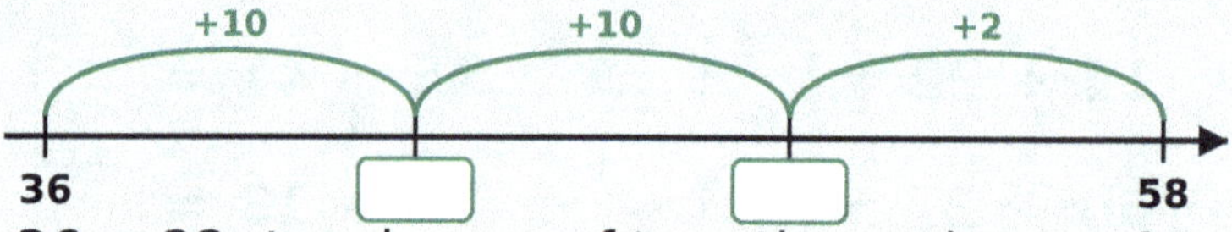

36 + 22: two jumps of ten, then a jump of 2.

WORKED EXAMPLE

Step 1. Find 47 + 25. Start at 47.

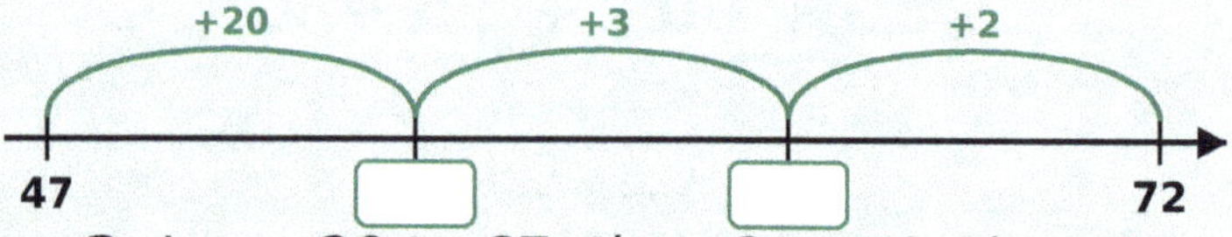

Step 2. Jump 20 to 67, then 3 to 70, then 2 more.

Answer: 72.

MATH TALK

Why do many people jump to the next ten before finishing?

TRY IT TOGETHER

Fill in the jumps and the answer.

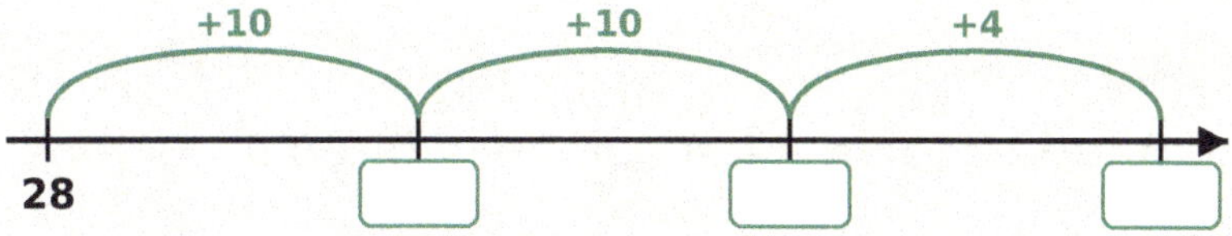

28 + 24 = ☐

ON YOUR OWN

Use the open number line to add.

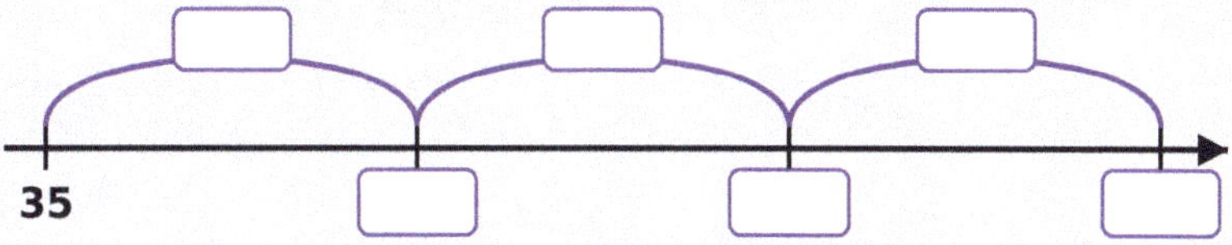

35 + 27 = ☐

Add.

1 **43 + 26 = ☐**

2 **58 + 31 = ☐**

3 **27 + 45 = ☐**

4 **64 + 18 = ☐**

3 Breaking Numbers Apart

Standard: 2.NBT.B.5

NEW WORDS

expanded form — a number written as its tens and ones

LET'S LEARN

Break both numbers into tens and ones, add the parts, then put them back together.

$$58 = 50 + 8$$

$$26 = 20 + 6$$

$$70 + 14 = 84$$

WORKED EXAMPLE

Step 1. 37 + 45. Break both: 30 + 7 and 40 + 5.

Step 2. Tens: 30 + 40 = 70. Ones: 7 + 5 = 12.

$$70 + 12 = 82$$

Answer: 82.

MATH TALK

In the example, the ones made 12. What happens to the extra ten?

TRY IT TOGETHER

Break the numbers apart and add.

46 = ☐ + ☐

33 = ☐ + ☐

46 + 33 = ☐

ON YOUR OWN

Add by breaking the numbers apart.

1. 24 + 38 = ☐
2. 56 + 27 = ☐
3. 19 + 44 = ☐
4. 65 + 29 = ☐

5. Write 73 in expanded form.

73 = ☐ + ☐

4 Regrouping Ones

Standard: 2.NBT.B.5

NEW WORDS

regroup — to trade ten ones for one ten

LET'S LEARN

When the ones make 10 or more, trade ten ones for one ten.

	10s	1s
	2	8
+	1	5
	4	3

$8 + 5 = 13$ ones. Trade ten of them for one ten, and 3 ones stay.

WORKED EXAMPLE

Step 1. Add the ones: $6 + 7 = 13$.

	10s	1s
	4	6
+	2	7
	7	3

Step 2. Write 3 in the ones and carry 1 ten.
Step 3. Add the tens: $1 + 4 + 2 = 7$.
Answer: 73.

MATH TALK

How do you know when you need to regroup before you start?

TRY IT TOGETHER

Add. Regroup if you need to.

10s	1s
3	5
+ 2	8
☐	☐

10s	1s
4	7
+ 1	6
☐	☐

ON YOUR OWN

Add.

5	8
+ 2	4
☐	☐

3	6
+ 4	7
☐	☐

2	9
+ 6	5
☐	☐

1 $48 + 37 = \square$

2 $53 + 29 = \square$

REMEMBER

If the ones digits add to less than 10, no regrouping is needed.

5 Adding Several Numbers

Standard: 2.NBT.B.6

NEW WORDS

total — the sum of all the numbers

LET'S LEARN

Add the tens of every number, then all the ones, then join the two parts.

$$23 + 14 + 32$$

$$20 + 10 + 30 = 60$$

$$3 + 4 + 2 = 9$$

$$60 + 9 = 69$$

WORKED EXAMPLE

Step 1. Add 15 + 25 + 12. Tens: 10 + 20 + 10 = 40.

Step 2. Ones: 5 + 5 + 2 = 12.

$$40 + 12 = 52$$

Answer: 52.

MATH TALK

Would it help to look for two ones digits that make ten? Show where that happens in the example.

TRY IT TOGETHER

Find the total.

1 **21 + 13 + 34 =** ☐

2 **16 + 24 + 11 =** ☐

ON YOUR OWN

Add.

1 **12 + 23 + 31 =** ☐

2 **18 + 22 + 15 =** ☐

3 **25 + 14 + 26 =** ☐

4 **11 + 22 + 13 + 14 =** ☐

5. Four boxes hold 21, 15, 30, and 12 books. How many books in all? ______

Math in My World: The Book Fair

Problem solving with Unit 3 skills

The school book fair is counting sales.

THE GRAPH

Books sold

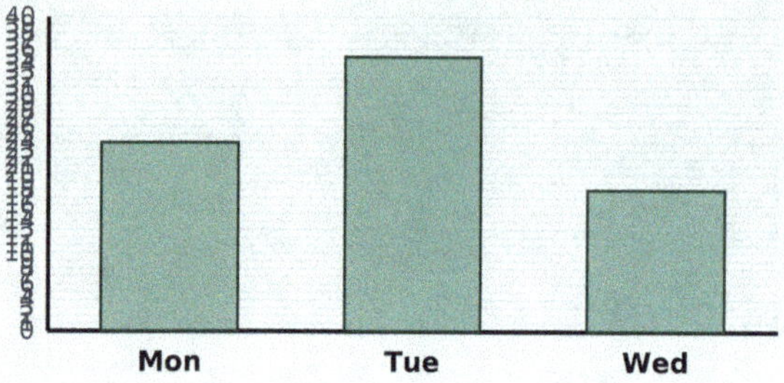

YOUR TURN

1. How many books sold on Tuesday? ______
2. How many books sold on Monday and Tuesday together?

☐ + ☐ = ☐

3. How many books sold in the three days? ______
4. On Thursday 26 more books sold. What is the new three-day total for Tuesday, Wednesday, and Thursday? ______
5. Did more books sell on Monday or on Wednesday? By how many?

MATH TALK

For question 3 you added three two-digit numbers. Describe the order you added them in and why.

Unit 3 Review

Review of Lessons 3.1–3.5

A Part A · Add tens and ones

1 $34 + 25 = \square$

2 $61 + 27 = \square$

3 $42 + 36 = \square$

4 $13 + 55 = \square$

B Part B · Open number line

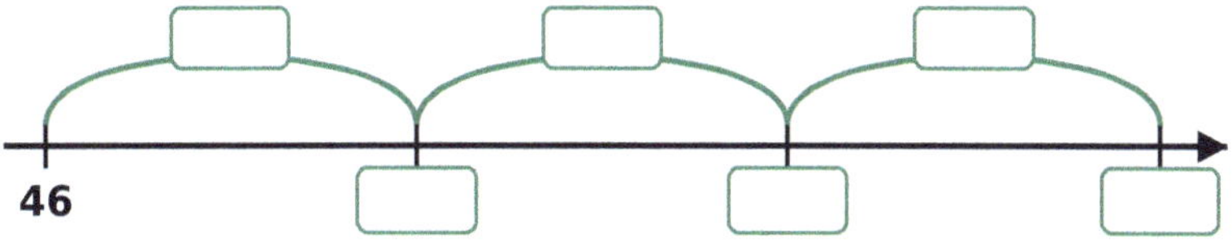

$46 + 27 = \square$

C Part C · Regroup

10s	1s
☐	☐
3	8
+ 2	6
☐	☐

10s	1s
☐	☐
5	7
+ 3	5
☐	☐

D Part D · Expanded form

68 = ☐ + ☐

E Part E · Several numbers

5 **14 + 21 + 32 =** ☐

6 **23 + 17 + 20 =** ☐

F Part F · Word problem

7. A class read 38 books in May and 47 books in June. How many books did the class read altogether?

CHALLENGE

Find two two-digit numbers that add to exactly 80, where adding them needs regrouping. Show your work.

UNIT

4

Subtracting Two-Digit Numbers

YOUR LEARNING GOALS

In this unit you will learn to...

- Subtract tens from tens and ones from ones.
- Count back or count up to find a difference.
- Regroup one ten into ten ones.
- Check a subtraction by adding.
- Solve one- and two-step word problems.

STANDARDS COVERED

2.NBT.B.5 Fluently add and subtract within 100 using strategies based on place value.

2.OA.A.1 Use addition and subtraction within 100 to solve one- and two-step word problems.

2.NBT.B.9 Explain why addition and subtraction strategies work.

1 Subtracting Tens and Ones

Standard: 2.NBT.B.5

NEW WORDS

difference — the answer when you subtract

LET'S LEARN

Take away the tens, then take away the ones.

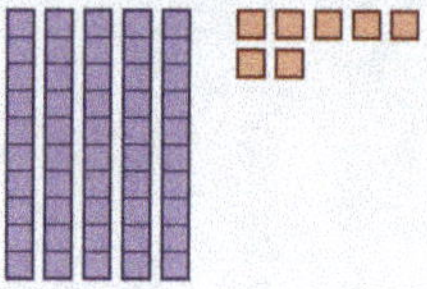

57 − 24

50 − 20 = 30

7 − 4 = 3

30 + 3 = 33

WORKED EXAMPLE

Step 1. Find 86 − 32. Tens: 80 − 30 = 50.

Step 2. Ones: 6 − 2 = 4.

50 + 4 = 54

Answer: 54.

MATH TALK

Why does this method only work when the top ones digit is big enough?

TRY IT TOGETHER

Subtract the tens, then the ones.

1 **68 − 35 =** ☐ 2 **79 − 42 =** ☐

ON YOUR OWN

Subtract.

1 **47 − 23 =** ☐ 2 **95 − 51 =** ☐

3 **88 − 40 =** ☐ 4 **56 − 34 =** ☐

5 **73 − 62 =** ☐ 6 **99 − 27 =** ☐

7. A shelf held 65 books. 31 were borrowed.
How many are left? ______

2 Counting Back and Counting Up

Standard: 2.NBT.B.5

NEW WORDS

count up — to start at the smaller number and jump forward

LET'S LEARN

You can subtract by counting **back**, or by counting **up** from the smaller number.

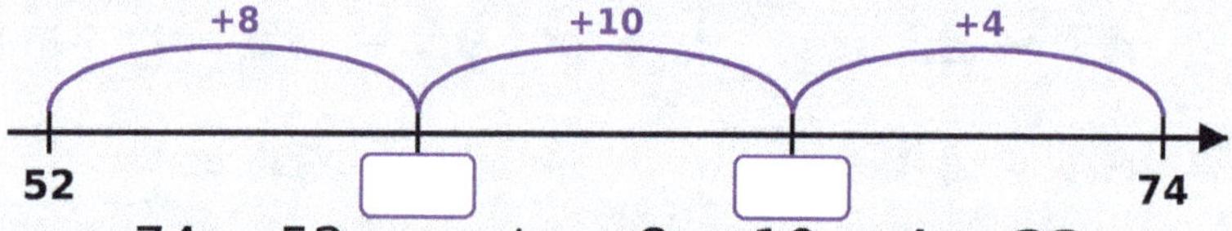

74 − 52: count up 8 + 10 + 4 = **22**.

WORKED EXAMPLE

Step 1. Find 63 − 28 by counting up from 28.

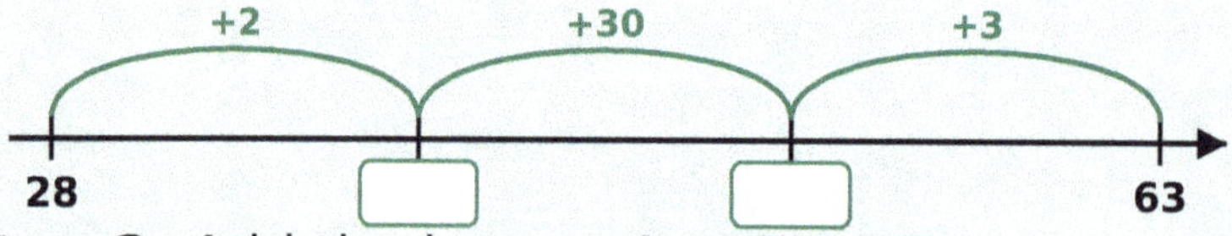

Step 2. Add the jumps: 2 + 30 + 3.

Answer: 35.

MATH TALK

When would counting up be faster than counting back?

TRY IT TOGETHER

Count up to find the difference.

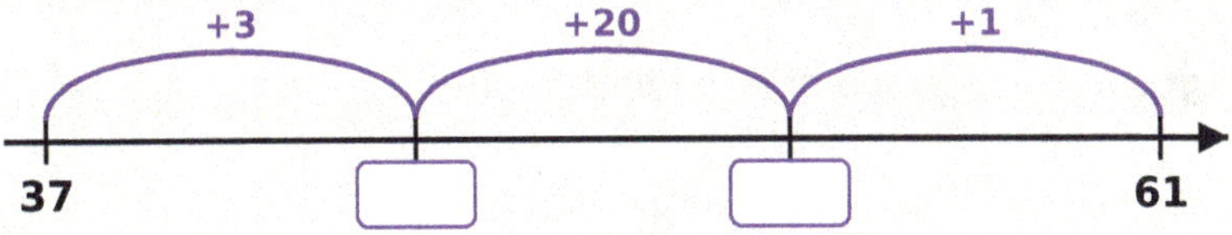

61 – 37 = ☐

ON YOUR OWN

Use the open number line.

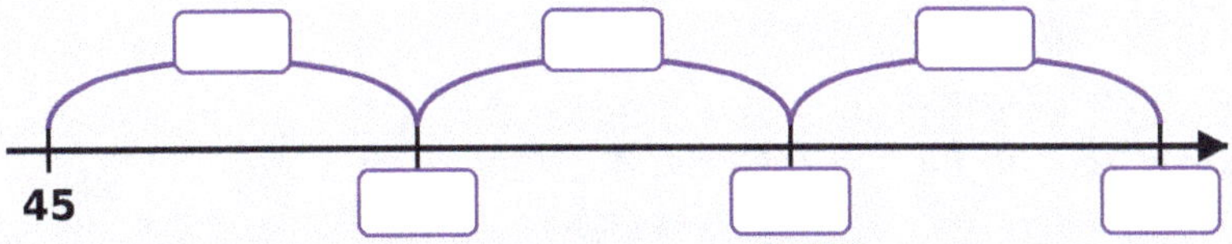

72 – 45 = ☐

Subtract.

1 **54 – 29 = ☐**

2 **81 – 36 = ☐**

3 **66 – 48 = ☐**

4 **90 – 55 = ☐**

3 Regrouping a Ten

Standard: 2.NBT.B.5

NEW WORDS

regroup — to trade one ten for ten ones

LET'S LEARN

If there are not enough ones, open up one ten into ten ones.

10s	1s
☐	☐
5	2
− 2	7
2	5

2 ones is not enough to take 7 away. Trade one ten, so the ones become 12.

WORKED EXAMPLE

Step 1. Find 64 − 38. Ones: 4 is less than 8.

10s	1s
☐	☐
6	4
− 3	8
2	6

Step 2. Trade: 6 tens 4 ones becomes 5 tens 14 ones.
Step 3. 14 − 8 = 6, and 5 − 3 = 2.
Answer: 26.

MATH TALK

After regrouping, the number still has the same value. Why is that true?

TRY IT TOGETHER

Subtract. Regroup if you need to.

10s	1s
4	3
− 1	7
☐	☐

10s	1s
7	1
− 2	6
☐	☐

ON YOUR OWN

Subtract.

$$\begin{array}{r} 85 \\ -\,39 \\ \hline \square\square \end{array} \qquad \begin{array}{r} 60 \\ -\,24 \\ \hline \square\square \end{array} \qquad \begin{array}{r} 92 \\ -\,57 \\ \hline \square\square \end{array}$$

1 $74 - 58 = \square$ 2 $53 - 19 = \square$

REMEMBER

Compare the ones digits before you begin.
Top digit smaller means you will regroup.

4 Checking with Addition

Standard: 2.NBT.B.9

NEW WORDS

check — to test an answer another way

LET'S LEARN

Addition undoes subtraction. Add your answer to the number you took away.

$$72 - 35 = 37$$

$$37 + 35 = 72$$

The check gives back the number you started with, so the answer is right.

WORKED EXAMPLE

Step 1. A student writes 81 − 46 = 45.

Step 2. Check: 45 + 46 = 91, not 81.

$$81 - 46 = 35$$

Answer: The correct difference is 35.

MATH TALK

If the check does not give back the first number, what should you do next?

TRY IT TOGETHER

Subtract, then check by adding.

1 **64 − 27 = ☐ check: ☐ + 27 = ☐**

2 **50 − 18 = ☐ check: ☐ + 18 = ☐**

ON YOUR OWN

Is each answer right? Write **yes** or **no**.

1. 45 − 18 = 27 ______
2. 83 − 29 = 64 ______
3. 70 − 36 = 34 ______

Subtract and check.

4 **91 − 47 = ☐**   5 **62 − 38 = ☐**

5 Word Problems within 100

Standard: 2.OA.A.1

NEW WORDS

two-step problem — a problem that needs two calculations

LET'S LEARN

Draw a bar model. Decide whether you know the whole or the parts.

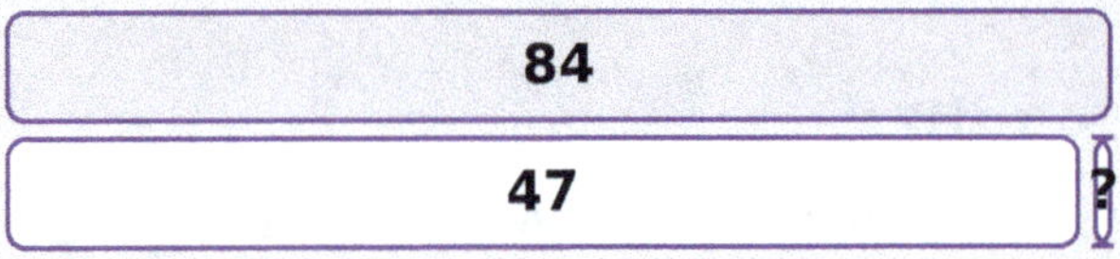

A whole of 84 with one part 47: subtract to find the other part.

$$84 - 47 = 37$$

WORKED EXAMPLE

Step 1. A shop had 56 apples. It sold 28, then got 15 more.

$$56 - 28 = 28$$

Step 2. Now add the delivery.

$$28 + 15 = 43$$

Answer: 43 apples.

MATH TALK

How can you tell that the apple problem needs two steps?

TRY IT TOGETHER

1. There are 72 seats. 45 are taken. How many are empty?

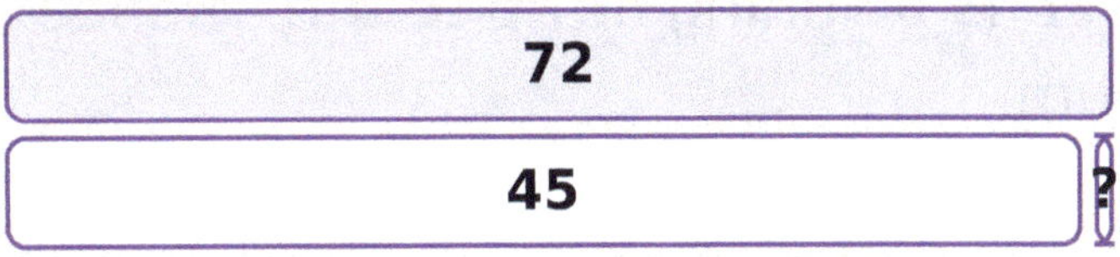

$\square - \square = \square$

ON YOUR OWN

2. Nia had 38 stickers. She was given 26 more, then used 19. How many does she have now?

3. A bus had 40 riders. 17 got off and 12 got on. How many riders are on the bus? ______

4. Tom read 24 pages on Monday and 31 on Tuesday. How many more pages must he read to reach 75? ______

Math in My World: The Class Trip

Problem solving with Unit 4 skills

Class 2B is planning a trip to the science museum.

THE INFORMATION

Trip facts

Seats on the bus: **52**

Children going: **28**

Adults going: **9**

Museum tickets already bought: **34**

YOUR TURN

1. How many people are going in all? ______
2. How many bus seats will be empty? ______
3. How many more tickets are needed? ______
4. Another class of 15 children joins the trip. How many people are going now? ______
5. Are there enough bus seats for everyone in question 4? __________

MATH TALK

Question 3 could be solved by subtracting or by counting up. Which did you use, and why?

Unit 4 Review

Review of Lessons 4.1–4.5

A Part A · Subtract

1 $68 - 24 = \square$

2 $95 - 43 = \square$

3 $57 - 31 = \square$

4 $80 - 50 = \square$

B Part B · Regroup

10s	1s
☐	☐
7	2
− 3	8
☐	☐

10s	1s
☐	☐
5	1
− 2	6
☐	☐

C Part C · Count up

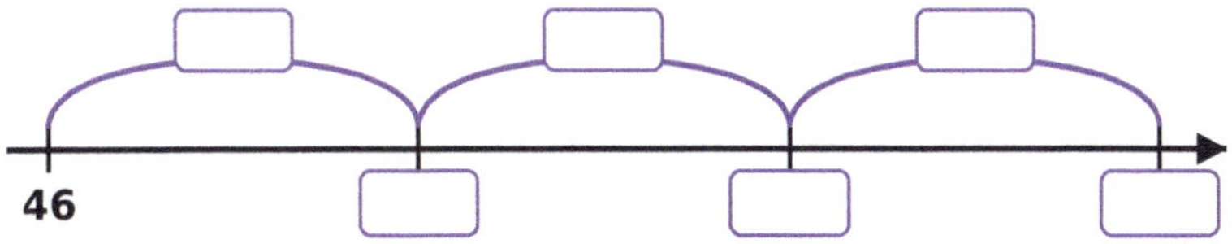

$73 - 46 = \square$

D Part D · Check by adding

Is 84 − 37 = 47 correct? Show your check.

E Part E · Word problems

5. A jar held 90 beads. 46 were used. How many beads are left? ______

6. A team scored 35 points, then 28 more, then lost 19. What is the score now? ______

CHALLENGE

Write a subtraction with two-digit numbers that needs regrouping and has the answer 26. Show your check.

UNIT

5

Numbers to 1,000

YOUR LEARNING GOALS

In this unit you will learn to...

- Read a three-digit number as hundreds, tens, and ones.
- Write numbers to 1,000 in numerals and in words.
- Write a number in expanded form.
- Skip-count by 5s, 10s, and 100s.
- Compare three-digit numbers with >, =, and <.

STANDARDS COVERED

2.NBT.A.1 Understand that the three digits of a three-digit number represent hundreds, tens, and ones.

2.NBT.A.2 Count within 1000; skip-count by 5s, 10s, and 100s.

2.NBT.A.3 Read and write numbers to 1000 using numerals, number names, and expanded form.

2.NBT.A.4 Compare two three-digit numbers using >, =, and <.

1 Hundreds, Tens, and Ones

Standard: 2.NBT.A.1

NEW WORDS

hundred — ten groups of ten

flat — a block that shows one hundred

LET'S LEARN

Ten tens make one **hundred**. A flat block shows 100.

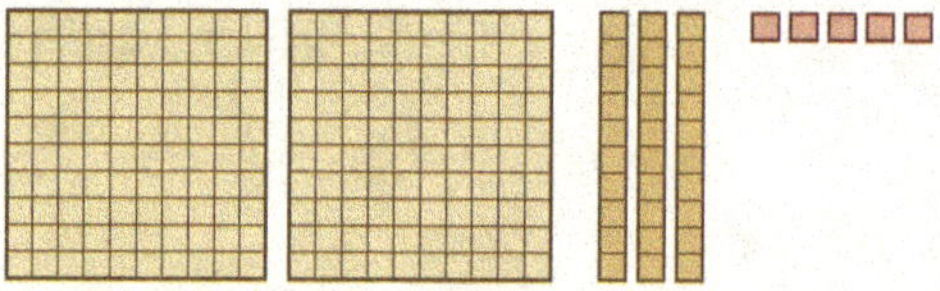

Hundreds	Tens	Ones
2	3	5

2 hundreds, 3 tens, and 5 ones is **235**.

WORKED EXAMPLE

Step 1. Count the flats: 4 hundreds is 400.

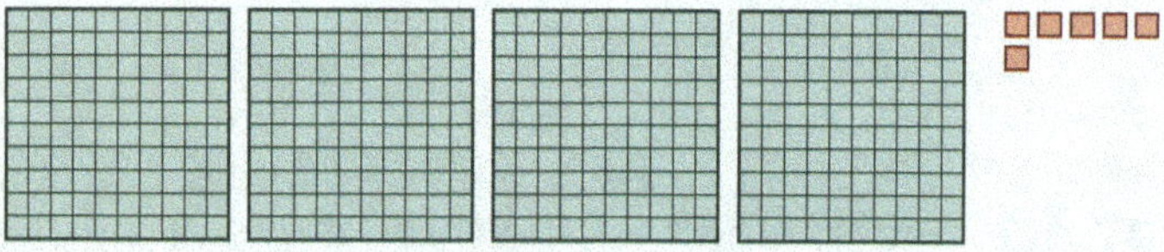

Step 2. There are no tens, and 6 ones.

Answer: 406. The zero holds the tens place.

MATH TALK

Why can't we just write 46 instead of 406?

TRY IT TOGETHER

Write the number shown.

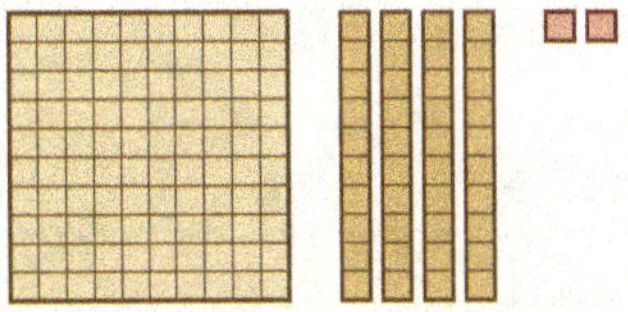

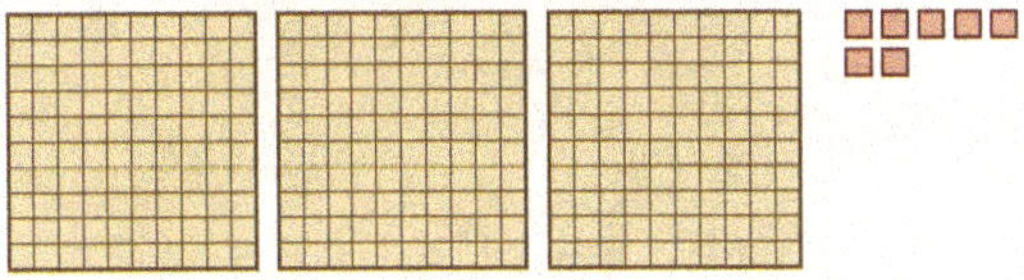

ON YOUR OWN

Write each number as hundreds, tens, and ones.

1 **472 = ☐ h ☐ t ☐ o**

2 **308 = ☐ h ☐ t ☐ o**

3 **690 = ☐ h ☐ t ☐ o**

Write the number.

4 **5 hundreds 1 ten 3 ones = ☐**

5 **8 hundreds 0 tens 9 ones = ☐**

REMEMBER

A zero in a place means there are none of that unit, but the zero still has to be written.

2 Reading and Writing Numbers

Standard: 2.NBT.A.3

NEW WORDS

number name — a number written in words

LET'S LEARN

Say the hundreds, then the tens and ones.

Hundreds	Tens	Ones
3	**4**	**7**

347 is read **three hundred forty-seven**.

Numbers from 21 to 99 use a hyphen: forty-seven, sixty-two.

WORKED EXAMPLE

Step 1. Write *five hundred sixteen* as a numeral.

Step 2. Five hundreds, one ten, six ones.

Hundreds	Tens	Ones
5	**1**	**6**

Answer: 516.

MATH TALK

How do you say a number like 205 aloud? Which place needs care?

TRY IT TOGETHER

Write each number in words.

1. 264 ____________________________

2. 730 ____________________________

ON YOUR OWN

Write the numeral.

1. four hundred eighty-one ______

2. nine hundred six ______

3. one hundred fifty ______

Write in words.

4. 375 ____________________________

5. 608 ____________________________

3 Expanded Form

Standard: 2.NBT.A.3

NEW WORDS

expanded form — a number written as the sum of its place values

LET'S LEARN

Expanded form shows what each digit is worth.

$$628 = 600 + 20 + 8$$

The 6 is worth 600, the 2 is worth 20, and the 8 is worth 8.

WORKED EXAMPLE

Step 1. Expand 405.

$$405 = 400 + 0 + 5$$

Step 2. The tens are zero, so that part may be left out.

Answer: 400 + 5.

MATH TALK

Expanded form and the place-value chart show the same thing. How are they different?

TRY IT TOGETHER

Write the expanded form.

573 = ☐ + ☐ + ☐

246 = ☐ + ☐ + ☐

ON YOUR OWN

Write the expanded form.

1. 812 = ______ + ______ + ______
2. 359 = ______ + ______ + ______
3. 704 = ______ + ______

Write the number.

4 **300 + 60 + 2 =** ☐

5 **500 + 9 =** ☐

6 **900 + 40 =** ☐

4 Skip-Counting to 1,000

Standard: 2.NBT.A.2

NEW WORDS

skip-count — to count in equal jumps

LET'S LEARN

Counting by hundreds changes only the hundreds digit.

100	200	300		

Counting by tens changes only the tens digit — until a new hundred is reached.

370	380	390		

WORKED EXAMPLE

Step 1. Count by tens from 480.

480	490	500	510

Step 2. After 490 the tens fill a new hundred.
Answer: 490, 500, 510.

MATH TALK

What happens to the digits when you count by tens past 590?

TRY IT TOGETHER

Fill in the missing numbers.

250		350		450
615	620		630	

ON YOUR OWN

Count by hundreds.

300			600	

Count by tens.

740			770	

Count by fives.

185			200	

1 **100 more than 428 is** ☐

2 **10 less than 306 is** ☐

5 Comparing to 1,000

Standard: 2.NBT.A.4

NEW WORDS

compare — to decide which number is greater

LET'S LEARN

Compare hundreds first, then tens, then ones.

Hundreds	Tens	Ones
4	6	2

Hundreds	Tens	Ones
4	3	9

Both have 4 hundreds. 6 tens beats 3 tens, so $462 > 439$.

WORKED EXAMPLE

Step 1. Compare 517 and 571. Hundreds match.

Step 2. Tens: 1 is less than 7.

$$517 < 571$$

Answer: $517 < 571$.

MATH TALK

Is a longer number always greater? Test your idea with 99 and 100.

TRY IT TOGETHER

Write >, =, or <.

1 **382** **328**

2 **605** **650**

3 **247** **247**

4 **790** **709**

ON YOUR OWN

Write >, =, or <.

1 **418** **481**

2 **936** **639**

3 **500** **500**

4 **273** **271**

5 **160** **106**

6 **845** **854**

Order from least to greatest: 528, 285, 582

Math in My World: The Recycling Drive

Problem solving with Unit 5 skills

Three classes collected cans for recycling.

THE INFORMATION

Cans collected

Class 2A: **246**

Class 2B: **264**

Class 2C: **240**

YOUR TURN

1. Which class collected the most cans? ______
2. Write 264 in expanded form.

   264 = ☐ + ☐ + ☐

3. Write 240 in words. ____________________
4. Order the three numbers from least to greatest.

   ☐ ☐ ☐

5. Class 2C collects 10 more cans. What is their new total? ______

MATH TALK

246 and 264 use the same three digits. Explain how you decided which one is greater.

Unit 5 Review

Review of Lessons 5.1–5.5

A Part A · Place value

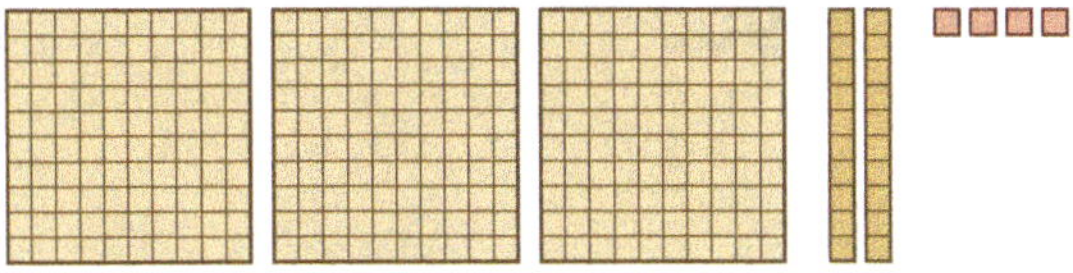

1 586 = ☐ h ☐ t ☐ o

B Part B · Expanded form

417 = ☐ + ☐ + ☐

2 600 + 30 + 9 = ☐

C Part C · Number names

3. Write 275 in words.

____

4. Write *eight hundred four* as a numeral. ______

D Part D · Skip-count

420			450	

700		900	

E Part E · Compare

5 319 ☐ 391

6 560 ☐ 506

7 748 ☐ 748

8 203 ☐ 230

CHALLENGE

Use the digits 3, 7, and 5 once each. Write the greatest and the least three-digit number you can make, then compare them with a symbol.

UNIT

6

Adding and Subtracting to 1,000

YOUR LEARNING GOALS

In this unit you will learn to...

- Add and subtract 10 or 100 in your head.
- Add hundreds and tens with blocks.
- Add three-digit numbers with regrouping.
- Subtract three-digit numbers with regrouping.
- Explain why an addition or subtraction method works.

STANDARDS COVERED

2.NBT.B.7 Add and subtract within 1000 using models or drawings and strategies based on place value.

2.NBT.B.8 Mentally add 10 or 100 to, and subtract 10 or 100 from, a given number 100–900.

2.NBT.B.9 Explain why addition and subtraction strategies work.

1 Ten and Hundred More or Less

Standard: 2.NBT.B.8

NEW WORDS

mentally — in your head, without writing

LET'S LEARN

Adding 10 changes only the tens digit. Adding 100 changes only the hundreds digit.

Hundreds	Tens	Ones
3	4	6

$$346 + 10 = 356$$

$$346 + 100 = 446$$

WORKED EXAMPLE

Step 1. Find 100 less than 528.

Step 2. Take one hundred away: 5 hundreds becomes 4.

$$528 - 100 = 428$$

Answer: 428.

MATH TALK

Which digits stay the same when you add 100 to a number?

TRY IT TOGETHER

Answer in your head.

1. **10 more than 273 is** ☐
2. **100 more than 481 is** ☐
3. **10 less than 650 is** ☐
4. **100 less than 739 is** ☐

ON YOUR OWN

Add or subtract mentally.

1. **165 + 10 =** ☐
2. **402 + 100 =** ☐
3. **594 − 10 =** ☐
4. **836 − 100 =** ☐
5. **297 + 10 =** ☐
6. **703 − 100 =** ☐

7. Start at 245. Add 100 three times. Where do you land? ______

REMEMBER

Watch for a carry: 297 + 10 is 307, because 9 tens plus 1 ten fills a new hundred.

2 Adding Hundreds and Tens

Standard: 2.NBT.B.7

NEW WORDS

place value — what a digit is worth because of its place

LET'S LEARN

Add hundreds to hundreds, tens to tens, and ones to ones.

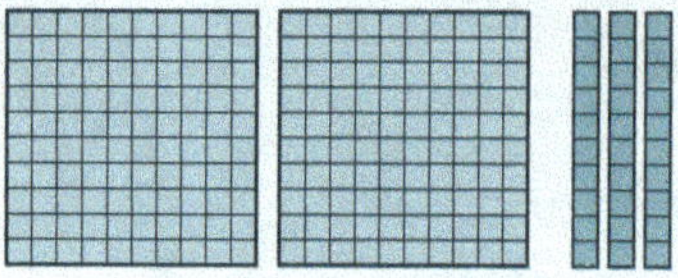

230 + 140

200 + 100 = 300

30 + 40 = 70

300 + 70 = 370

WORKED EXAMPLE

Step 1. Find 315 + 242. Hundreds: 300 + 200 = 500.

Step 2. Tens: 10 + 40 = 50. Ones: 5 + 2 = 7.

500 + 50 + 7 = 557

Answer: 557.

MATH TALK

Why does breaking a number into hundreds, tens, and ones make adding easier?

TRY IT TOGETHER

Add by place value.

1. 123 + 245 = ☐
2. 406 + 182 = ☐

ON YOUR OWN

Add.

1. 214 + 153 = ☐
2. 320 + 460 = ☐
3. 507 + 231 = ☐
4. 145 + 603 = ☐
5. 260 + 310 = ☐
6. 432 + 255 = ☐

7. Write 476 in expanded form, then add 200.

476 = ☐ + ☐ + ☐

3 Adding with Regrouping

Standard: 2.NBT.B.7

NEW WORDS

carry — to move a new ten or hundred to the next place

LET'S LEARN

When a place makes 10 or more, trade it for one of the next place.

$$\begin{array}{rccc} & \text{100s} & \text{10s} & \text{1s} \\ & 2 & 6 & 8 \\ + & 1 & 5 & 7 \\ \hline & 4 & 2 & 5 \end{array}$$

8 + 7 = 15 ones: write 5, carry 1 ten. Then 1 + 6 + 5 = 12 tens: write 2, carry 1 hundred.

WORKED EXAMPLE

Step 1. Find 346 + 275. Ones: 6 + 5 = 11.

$$\begin{array}{rccc} & \text{100s} & \text{10s} & \text{1s} \\ & 3 & 4 & 6 \\ + & 2 & 7 & 5 \\ \hline & 6 & 2 & 1 \end{array}$$

Step 2. Tens: 1 + 4 + 7 = 12. Write 2, carry 1.
Step 3. Hundreds: 1 + 3 + 2 = 6.
Answer: 621.

MATH TALK

A carried 1 above the tens column is worth how much? Explain.

TRY IT TOGETHER

Add. Regroup where you need to.

100s	10s	1s
☐	☐	☐
1	5	8
+ 2	3	4
☐	☐	☐

100s	10s	1s
☐	☐	☐
4	2	7
+ 1	9	5
☐	☐	☐

ON YOUR OWN

Add.

☐	☐	☐
3	6	5
+ 2	4	8
☐	☐	☐

☐	☐	☐
1	8	4
+ 2	7	9
☐	☐	☐

1 247 + 168 = ☐

2 356 + 285 = ☐

4 Subtracting with Regrouping

Standard: 2.NBT.B.7

NEW WORDS

regroup — to open one hundred into ten tens, or one ten into ten ones

LET'S LEARN

If a place has too few, open up one unit from the place to its left.

$$\begin{array}{cccc} & \text{100s} & \text{10s} & \text{1s} \\ & 4 & 3 & 2 \\ - & 1 & 7 & 5 \\ \hline & \boxed{2} & \boxed{5} & \boxed{7} \end{array}$$

2 ones cannot take 5, so open a ten: 12 − 5 = 7.

WORKED EXAMPLE

Step 1. Find 605 − 238. Ones: 5 is less than 8.

$$\begin{array}{cccc} & \text{100s} & \text{10s} & \text{1s} \\ & 6 & 0 & 5 \\ - & 2 & 3 & 8 \\ \hline & \boxed{3} & \boxed{6} & \boxed{7} \end{array}$$

Step 2. There are no tens, so open a hundred first: 6 hundreds becomes 5 hundreds and 10 tens.
Step 3. Then open a ten: 15 − 8 = 7.
Answer: 367.

MATH TALK

Why is subtracting across a zero the trickiest case?

TRY IT TOGETHER

Subtract. Regroup where you need to.

100s	10s	1s
3	7	4
− 1	2	8
☐	☐	☐

100s	10s	1s
5	2	0
− 2	4	5
☐	☐	☐

ON YOUR OWN

Subtract.

4	6	3
− 1	8	7
☐	☐	☐

7	0	0
− 3	5	6
☐	☐	☐

1 **532 − 279 =** ☐

2 **845 − 168 =** ☐

REMEMBER

Check every subtraction by adding your answer to the number you took away.

5 Explaining Why It Works

Standard: 2.NBT.B.9

NEW WORDS

strategy — a planned way of working something out

LET'S LEARN

A good explanation names the places and says what was traded.

$$198 + 45 = 200 + 43$$

Giving 2 to the 198 makes a friendly 200. Take the same 2 from the 45 so the total does not change.

$$200 + 43 = 243$$

WORKED EXAMPLE

Step 1. Nina says 305 − 99 is the same as 306 − 100.

Step 2. She added 1 to both numbers, so the gap between them is unchanged.

$$306 - 100 = 206$$

Answer: 206. Nina's method works.

MATH TALK

Adding the same amount to both numbers keeps a subtraction the same. Why?

TRY IT TOGETHER

Make the first number friendly, then add.

1 $297 + 36 = 300 + \square = \square$

2 $199 + 64 = 200 + \square = \square$

ON YOUR OWN

1. Explain why 250 + 250 = 500 without adding in columns.

2. Leo says 460 − 198 is the same as 462 − 200. Is he right? Explain.

3. Use a friendly number to find 403 − 97.

Math in My World: The Library Count

Problem solving with Unit 6 skills

The town library is counting its books.

THE INFORMATION

Books on the shelves

Picture books: **248**

Story books: **376**

Fact books: **195**

YOUR TURN

1. How many picture books and story books together?

☐ + ☐ = ☐

2. How many more story books than fact books? _____
3. 100 more fact books arrive. How many fact books now? _____
4. How many books are there in all? _____
5. 150 books are lent out. How many are left on the shelves? _____

MATH TALK

For question 2 you could subtract, or make 195 friendly by calling it 200. Explain how the friendly number helps.

Unit 6 Review

Review of Lessons 6.1–6.5

A Part A · Mental math

1 **10 more than 462 is** ☐

2 **100 less than 815 is** ☐

B Part B · Add

	100s	10s	1s
	☐	☐	☐
	2	3	6
+	1	4	5
	☐	☐	☐

	100s	10s	1s
	☐	☐	☐
	3	7	8
+	2	6	4
	☐	☐	☐

C Part C · Add by place value

3 $305 + 241 =$ ☐

4 $420 + 360 =$ ☐

D Part D · Subtract

100s	10s	1s
5	4	1
− 2	6	7
☐	☐	☐

100s	10s	1s
6	0	0
− 3	4	8
☐	☐	☐

E Part E · Explain

5. Is 298 + 47 the same as 300 + 45? Explain.

F Part F · Word problem

6. A shop had 425 pencils. It sold 168 and then received 200 more. How many pencils does it have now? ______

CHALLENGE

Find a three-digit subtraction that needs regrouping in both the ones and the tens, with the answer 176. Show your check.

UNIT

7

Money

YOUR LEARNING GOALS

In this unit you will learn to...

- Name each coin and say what it is worth.
- Count a mixed group of coins.
- Use the dollar sign and the cent sign correctly.
- Make the same amount in more than one way.
- Solve word problems about money.

STANDARDS COVERED

2.MD.C.8 Solve word problems involving dollar bills, quarters, dimes, nickels, and pennies, using $ and ¢ symbols appropriately.

2.NBT.A.2 Skip-count by 5s, 10s, and 100s.

1 Coins and Their Values

Standard: 2.MD.C.8

NEW WORDS

cent — the smallest unit of money, written ¢

value — how much something is worth

LET'S LEARN

Each coin has its own value. Size does not tell you the value.

A nickel is bigger than a dime, but a dime is worth more.

WORKED EXAMPLE

Step 1. How much are 3 dimes worth?

10¢ 20¢ 30¢

Step 2. Count by tens: 10, 20, 30.

Answer: 30¢.

MATH TALK

Why is counting quarters harder than counting dimes?

TRY IT TOGETHER

Write the total value.

ON YOUR OWN

Write the value of each coin.

1 **penny =** ☐ **¢**

2 **nickel =** ☐ **¢**

3 **dime =** ☐ **¢**

4 **quarter =** ☐ **¢**

Write the total.

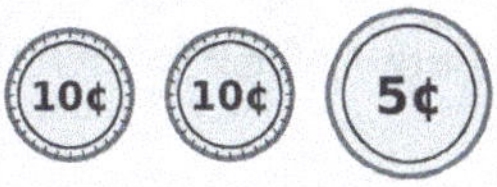

REMEMBER

Count the biggest coins first, then the smaller ones. That way you can skip-count for most of the total.

2 Counting Mixed Coins

Standard: 2.MD.C.8

NEW WORDS

skip-count — to count in equal jumps

LET'S LEARN

Sort the coins from greatest value to least, then count on.

25¢ 35¢ 45¢ 50¢ 51¢

25, 35, 45, 50, 51. The total is **51¢**.

WORKED EXAMPLE

Step 1. Sort: quarter, nickel, nickel, penny, penny.

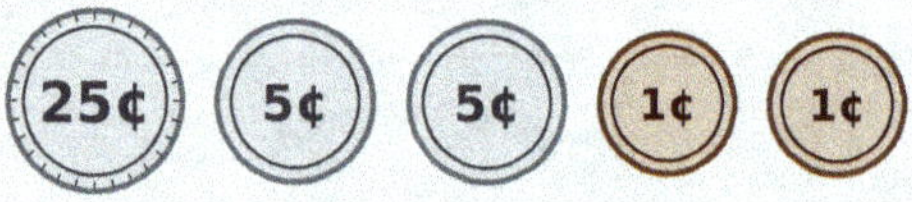

25¢ 30¢ 35¢ 36¢ 37¢

Step 2. Count: 25, 30, 35, 36, 37.

Answer: 37¢.

MATH TALK

Why is it easier to start with the quarter instead of the pennies?

TRY IT TOGETHER

Count the coins. Write the total.

ON YOUR OWN

Write the total value.

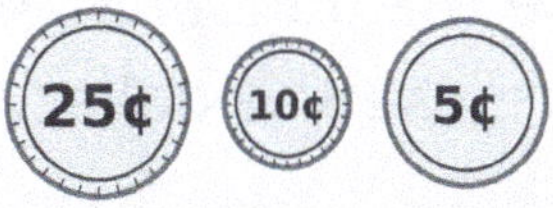

4. Which is worth more: 3 dimes or 5 nickels?

____

3 Dollars and Cents

Standard: 2.MD.C.8

NEW WORDS

dollar — one hundred cents, written $1
cent sign — the symbol ¢

LET'S LEARN

One hundred cents make one dollar.

one dollar

100 ¢ = $1

Amounts under a dollar use the cent sign: 45¢. Amounts of a dollar or more use the dollar sign: $2.

WORKED EXAMPLE

Step 1. How many quarters make one dollar?

25¢ **50¢** **75¢** **100¢**

Step 2. Count: 25, 50, 75, 100.
Answer: 4 quarters = $1.

MATH TALK

Why do we never write $45¢?

TRY IT TOGETHER

Write each amount with the correct symbol.

1 **sixty-three cents =**

2 **three dollars =**

ON YOUR OWN

How many of each coin make $1?

1 **pennies:** ☐ 2 **nickels:** ☐

3 **dimes:** ☐ 4 **quarters:** ☐

5. Write the amount.

☐

6. Is this more or less than one dollar?

____

4 Same Amount, Different Coins

Standard: 2.MD.C.8

NEW WORDS

equivalent — worth the same amount

LET'S LEARN

The same amount can be made with different coins.

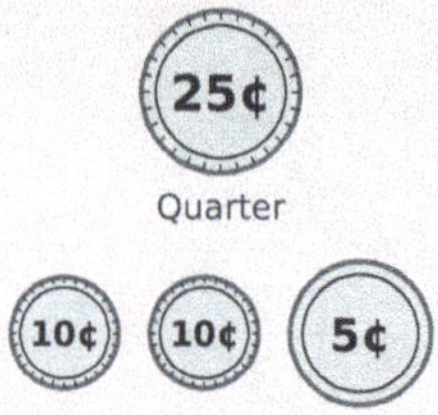

A quarter and 2 dimes with a nickel are both 25¢.

WORKED EXAMPLE

Step 1. Make 30¢ two ways.

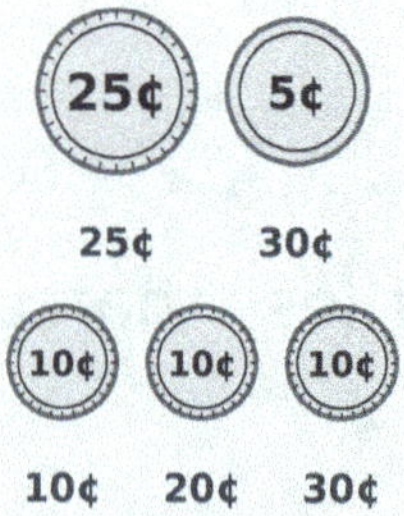

Answer: Both groups are worth 30¢.

MATH TALK

Which way of making 30¢ uses the fewest coins?

TRY IT TOGETHER

Draw or list a second way to make each amount.

= 35¢

ON YOUR OWN

Make each amount with the fewest coins. List the coins.

1. 40¢ ______________________________
2. 55¢ ______________________________
3. 68¢ ______________________________
4. Write two different ways to make 50¢.

REMEMBER

Fewest coins means starting with the largest coin that still fits inside the amount.

5 Money Word Problems

Standard: 2.MD.C.8

NEW WORDS

change — the money you get back

LET'S LEARN

Money problems work like any other addition or subtraction problem.

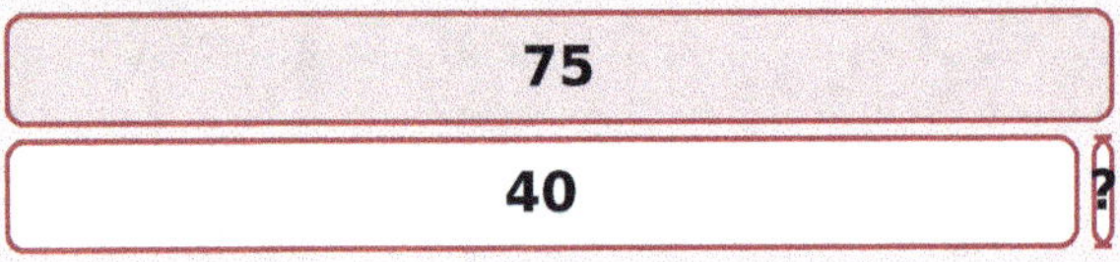

A toy costs 75¢. You have 40¢. Subtract to find how much more you need.

75 − 40 = 35

WORKED EXAMPLE

Step 1. A pencil costs 35¢. You pay with 2 quarters.

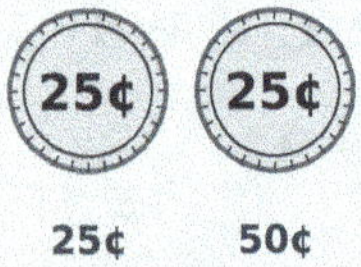

Step 2. You gave 50¢.

50 − 35 = 15

Answer: The change is 15¢.

MATH TALK

How could you check the change using coins instead of subtracting?

TRY IT TOGETHER

1. A sticker costs 28¢. You pay with 3 dimes. How much change do you get?

☐ – ☐ = ☐

ON YOUR OWN

2. An eraser costs 45¢ and a ruler costs 30¢. How much for both? ______
3. You have 80¢ and spend 55¢. How much is left? ______
4. A juice costs 65¢. You pay with 3 quarters. What is your change? ______
5. Ben has 2 quarters, 1 dime, and 3 pennies. Can he buy a book that costs 70¢? Explain.

Math in My World: The School Shop

Problem solving with Unit 7 skills

The school shop sells supplies at break time.

THE INFORMATION

Price list

Pencil **25¢** · Eraser **15¢** · Notebook **60¢** · Ruler **40¢**

What Ana has in her purse

YOUR TURN

1. How much money does Ana have? ______
2. Can she buy a notebook? __________
3. She buys a pencil and an eraser. How much does she spend?

   ☐ + ☐ = ☐

4. How much money does she have left?

   ______

5. What is the most expensive item she can still buy with the money left? __________

MATH TALK

Ana wants to pay for the ruler with the fewest coins. Which coins should she use?

Unit 7 Review

Review of Lessons 7.1–7.5

A Part A · Coin values

1 **nickel = ☐ ¢**

2 **quarter = ☐ ¢**

3 **dime = ☐ ¢**

4 **penny = ☐ ¢**

B Part B · Count the coins

☐

☐

C Part C · Dollars

5. How many dimes make $1? ______

6. Write forty-two cents with the correct symbol.

______

D Part D · Different ways

7. Make 45¢ with the fewest coins. List them.

E Part E · Word problems

8. A pen costs 55¢. You pay with 3 quarters. What is your change? ______

9. You have 90¢ and buy two items costing 35¢ and 25¢. How much is left? ______

CHALLENGE

Make 70¢ in three different ways. One way must use exactly six coins.

UNIT

8

Telling Time

YOUR LEARNING GOALS

In this unit you will learn to...

- Tell time to the nearest five minutes.
- Use quarter past and quarter to.
- Use a.m. and p.m. correctly.
- Match analog and digital clocks.
- Put the events of a day in time order.

STANDARDS COVERED

2.MD.C.7 Tell and write time from analog and digital clocks to the nearest five minutes, using a.m. and p.m.

2.NBT.A.2 Skip-count by 5s.

1 Time to Five Minutes

Standard: 2.MD.C.7

NEW WORDS

minute hand — the long hand
hour hand — the short hand

LET'S LEARN

Each number on the clock is five minutes apart. Count by fives around the dial.

2:20

The long hand is on 4, so count 5, 10, 15, 20 minutes.

WORKED EXAMPLE

Step 1. The short hand is just past 7.

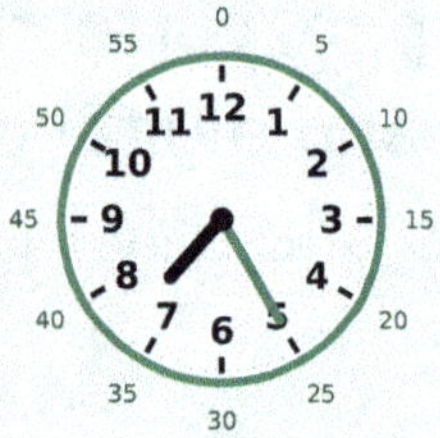

7:25

Step 2. The long hand is on 5, so count by fives to 25.
Answer: 7:25.

MATH TALK

Why does the hour hand move a little past the number as the minutes pass?

TRY IT TOGETHER

Write the time.

1 ☐ : ☐   2 ☐ : ☐

ON YOUR OWN

Write the time shown.

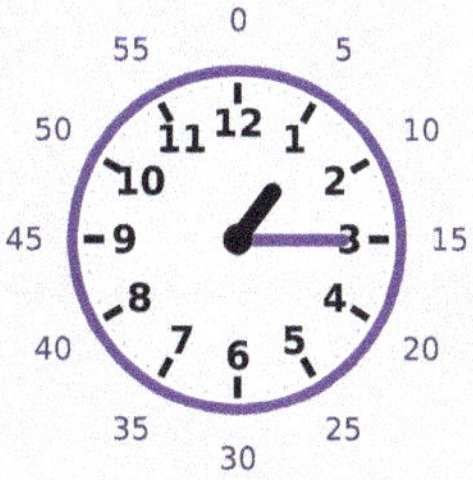

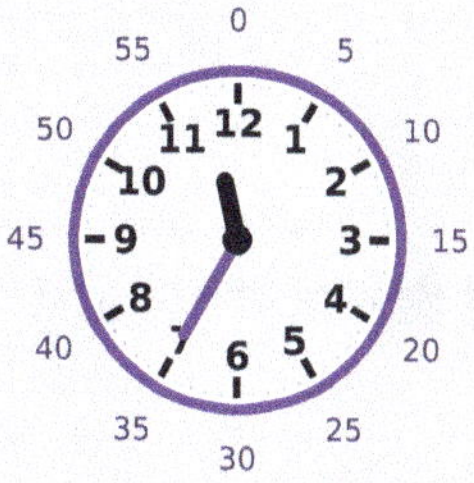

1 ☐ : ☐   2 ☐ : ☐   3 ☐ : ☐

4. Where does the long hand point at 4:45?

______

REMEMBER

Count by fives from the 12: each number is five more minutes.

2 Quarter Past and Quarter To

Standard: 2.MD.C.7

NEW WORDS

quarter past — fifteen minutes after the hour
quarter to — fifteen minutes before the next hour

LET'S LEARN

An hour has four quarters of fifteen minutes each.

quarter past 3

half past 3

quarter to 4

WORKED EXAMPLE

Step 1. The long hand is on 9.

6:45

Step 2. That is 45 minutes past 6, or 15 minutes before 7.
Answer: quarter to 7, written 6:45.

MATH TALK

Why does quarter to 7 show a 6 on a digital clock?

TRY IT TOGETHER

Write each time in words and in digits.

ON YOUR OWN

Match the time to the words.

1 **4:15** **2** **10:45** **3** **7:30**

a. half past 7 b. quarter past 4 c. quarter to 11

Draw the hands.

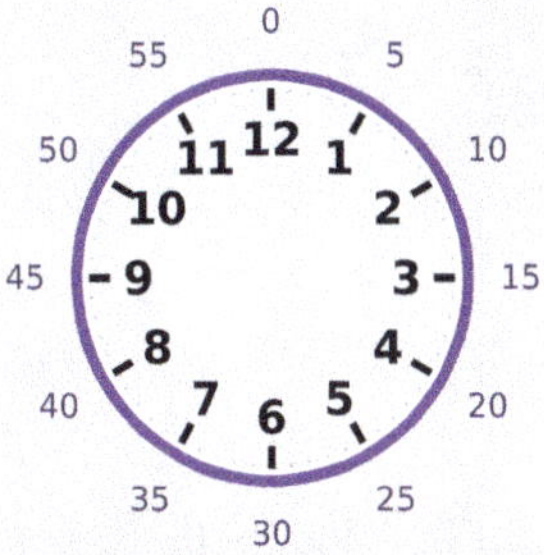

quarter past 5

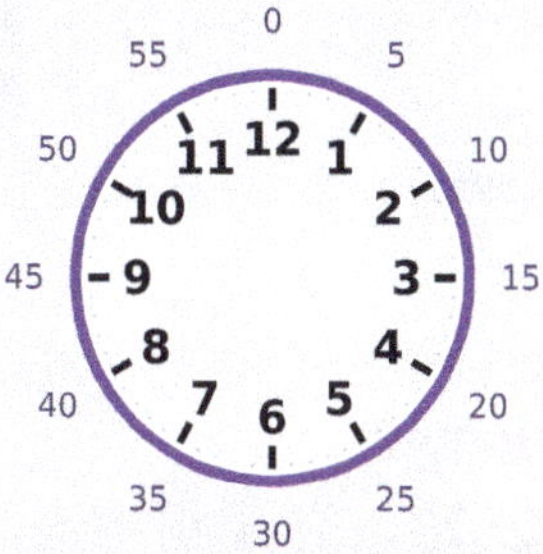

quarter to 9

3 A.M. and P.M.

Standard: 2.MD.C.7

NEW WORDS

a.m. — from midnight to noon
p.m. — from noon to midnight

LET'S LEARN

A clock face shows each time twice a day. **a.m.** and **p.m.** tell which one is meant.

7:30 a.m.

7:30 p.m.

Breakfast is at 7:30 a.m. Dinner is at 7:30 p.m.

WORKED EXAMPLE

Step 1. School starts in the morning.
Step 2. Morning times are before noon.

Answer: 8:30 a.m.

MATH TALK

Is 12:00 noon written a.m. or p.m.? Why is it confusing?

TRY IT TOGETHER

Write **a.m.** or **p.m.**

1. Waking up at 7:00 ______
2. Eating dinner at 6:30 ______
3. Going to bed at 8:45 ______

ON YOUR OWN

Write **a.m.** or **p.m.**

1. The sun rises at 6:15 ______
2. Recess at 10:30 ______
3. Watching the sunset at 7:45 ______
4. Brushing teeth before school at 7:20

 ______

5. Write one thing you do in the a.m. and one in the p.m.

4 Analog and Digital

Standard: 2.MD.C.7

NEW WORDS

analog clock — a clock with hands
digital clock — a clock that shows numbers

LET'S LEARN

Both clocks show the same time in different ways.

The digits before the colon are the hour. The digits after are the minutes.

WORKED EXAMPLE

Step 1. A digital clock shows 9:55.

9:55

Step 2. 55 minutes means the long hand is on 11.
Answer: Short hand almost at 10, long hand on 11.

MATH TALK

Which clock is easier to read quickly? Which tells you more about the shape of the hour?

TRY IT TOGETHER

Write the digital time.

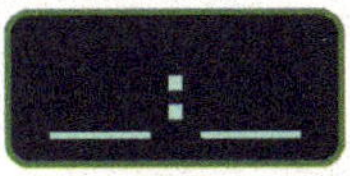

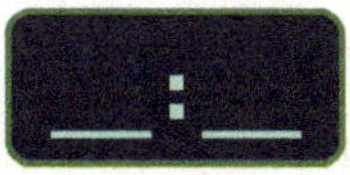

ON YOUR OWN

Write the digital time.

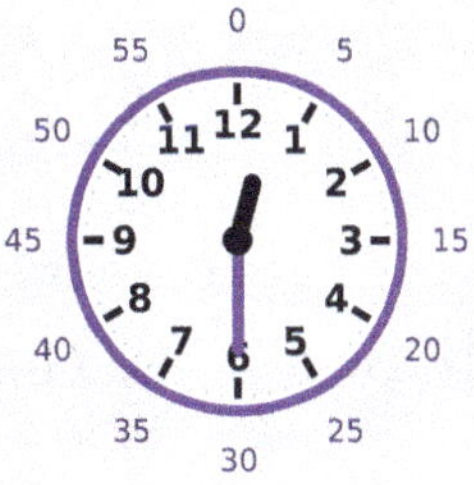

1 ☐ : ☐   2 ☐ : ☐   3 ☐ : ☐

Draw hands to show 3:40.

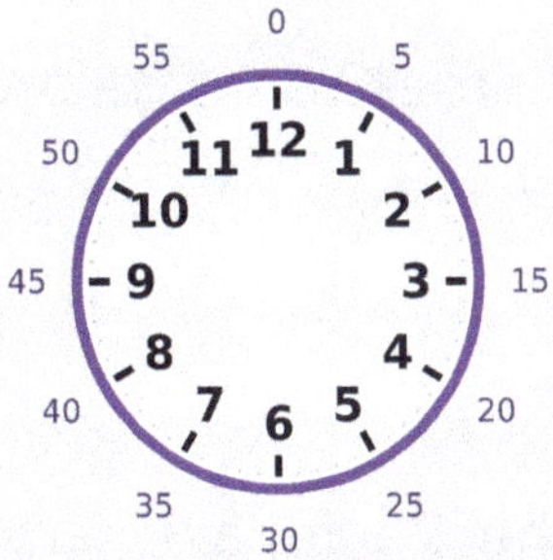

5 Ordering Times

Standard: 2.MD.C.7

NEW WORDS

earlier — before
later — after

LET'S LEARN

Compare the hour first. If the hours match, compare the minutes.

9:15 a.m.

9:45 a.m.

Both are 9 o'clock, but 15 minutes is earlier than 45.

WORKED EXAMPLE

Step 1. Put in order: 1:30 p.m., 8:00 a.m., 11:20 a.m.

Step 2. All a.m. times come before all p.m. times.

Answer: 8:00 a.m., 11:20 a.m., 1:30 p.m.

MATH TALK

Why must you check a.m. and p.m. before comparing the hours?

TRY IT TOGETHER

Circle the **earlier** time.

1 **3:15 p.m. or 3:50 p.m.**

2 **7:00 a.m. or 7:00 p.m.**

ON YOUR OWN

Put each set of times in order, earliest first.

1. 6:45 a.m., 6:15 a.m., 6:30 a.m.

2. 12:00 p.m., 9:30 a.m., 4:05 p.m.

3. Recess is at 10:15 a.m. Lunch is at 12:00 p.m. Which comes first? __________

4. Write a time that is later than 2:20 p.m. but earlier than 3:00 p.m. ______

Math in My World: My School Day

Problem solving with Unit 8 skills

Here is the timetable for Class 2B.

THE INFORMATION

Timetable

School starts **8:30 a.m.**

Reading **9:15 a.m.**

Recess **10:45 a.m.**

Lunch **12:15 p.m.**

Home time **3:00 p.m.**

YOUR TURN

1. What time does reading start? ______
2. Which happens first, recess or lunch? __________
3. Draw the hands to show recess time.

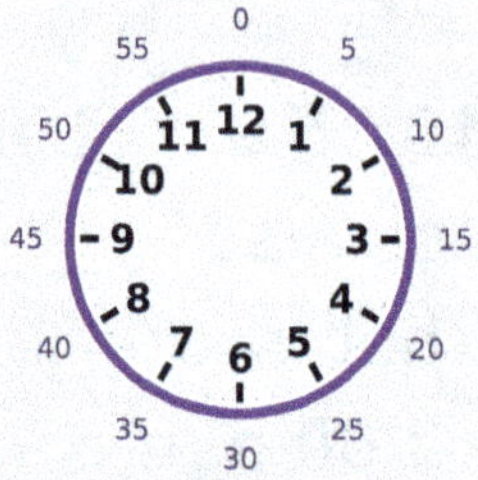

4. Is lunch in the a.m. or the p.m.? ______
5. Write home time in words.

____

MATH TALK

Two events on the timetable are quarter past the hour. Which are they, and how can you tell?

Unit 8 Review

Review of Lessons 8.1–8.5

A Part A · Read the clock

1 ☐ : ☐

2 ☐ : ☐

B Part B · Quarters

3. Write quarter past 6 in digits. ______

4. Write 8:45 in words. ____________________

C Part C · a.m. or p.m.

5. Eating breakfast at 7:15 ______

6. Playing outside after school at 4:00 ______

D Part D · Draw the hands

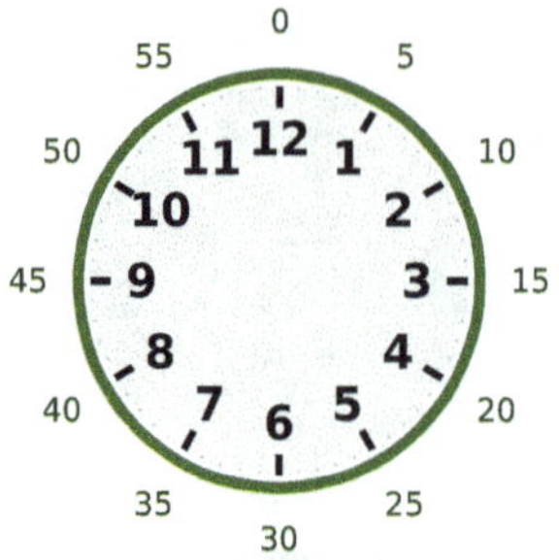

2:35

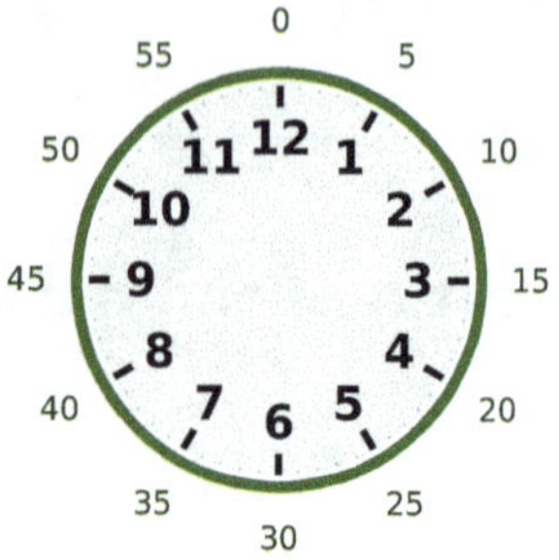

quarter to 5

E Part E · Order the times

7. Put in order, earliest first: 1:10 p.m., 8:40 a.m., 11:55 a.m.

CHALLENGE

A film starts at 2:15 p.m. and the long hand goes twice around the clock before it ends. What time does the film end?

UNIT

9

Measuring Length

YOUR LEARNING GOALS

In this unit you will learn to...

- Measure with a ruler in inches.
- Measure with a ruler in centimeters.
- Choose between inches, feet, centimeters, and meters.
- Estimate a length before measuring.
- Compare lengths and solve length problems.

STANDARDS COVERED

2.MD.A.1 Measure the length of an object using rulers, yardsticks, meter sticks, and measuring tapes.
2.MD.A.2 Measure the length of an object twice, using length units of different lengths.
2.MD.A.3 Estimate lengths using units of inches, feet, centimeters, and meters.
2.MD.A.4 Measure to determine how much longer one object is than another.
2.MD.B.5 Use addition and subtraction within 100 to solve word problems involving lengths.

1 Measuring in Inches

Standard: 2.MD.A.1

NEW WORDS

inch — a unit of length, written in.
ruler — a tool marked in equal units

LET'S LEARN

Line up one end of the object with the **0** mark. Read the number at the other end.

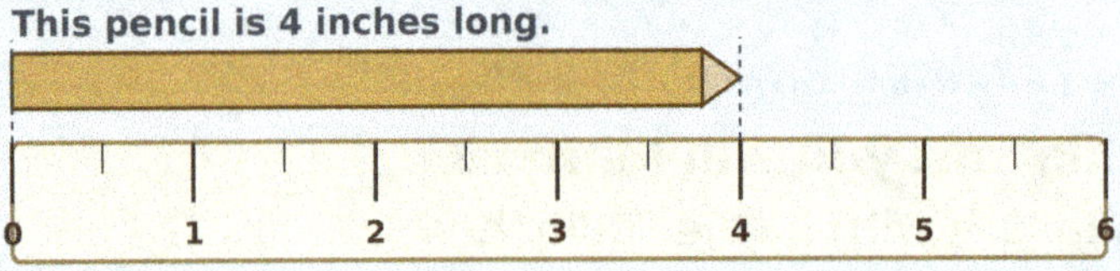

Always start at 0, not at the edge of the ruler.

WORKED EXAMPLE

Step 1. Put the left end of the crayon at 0.

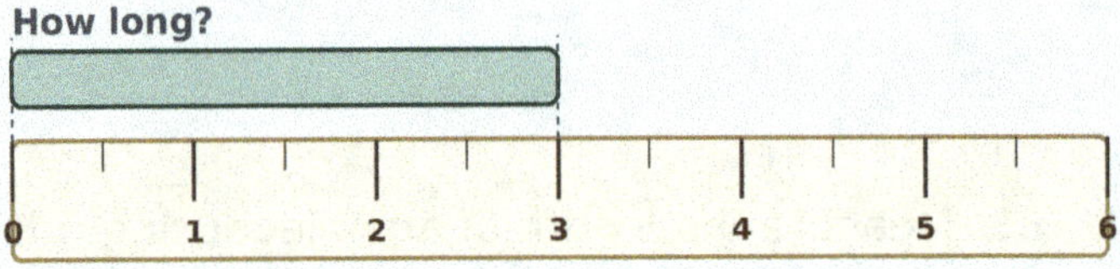

Step 2. The right end reaches 3.
Answer: 3 inches.

MATH TALK

What happens to your measurement if you start at 1 instead of 0?

TRY IT TOGETHER

Write the length in inches.

1. _____ inches

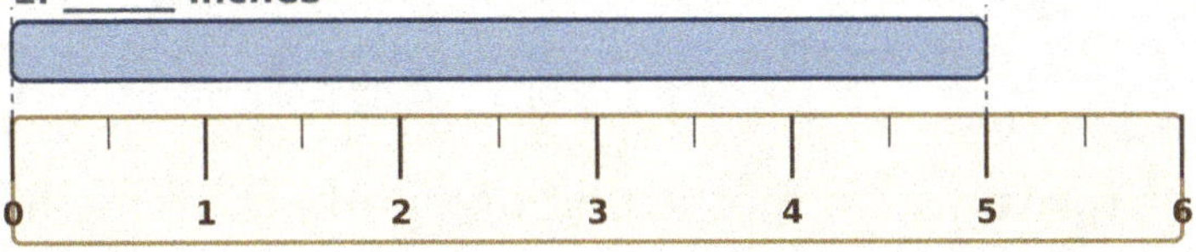

2. _____ inches

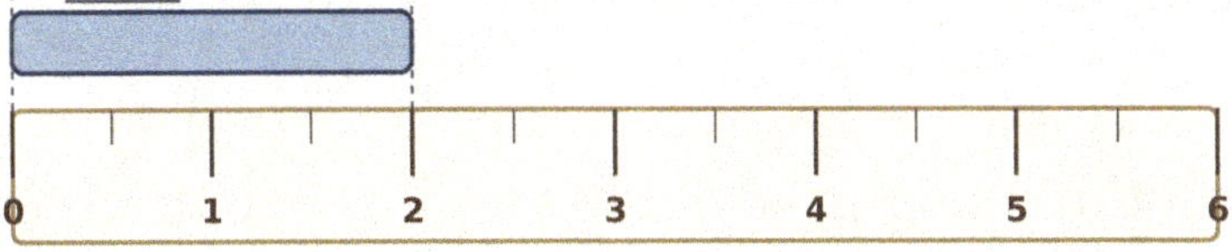

ON YOUR OWN

Write the length.

1. _____ inches

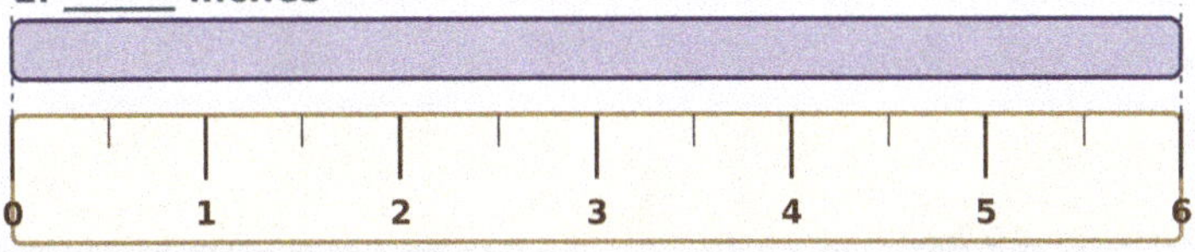

2. _____ inches

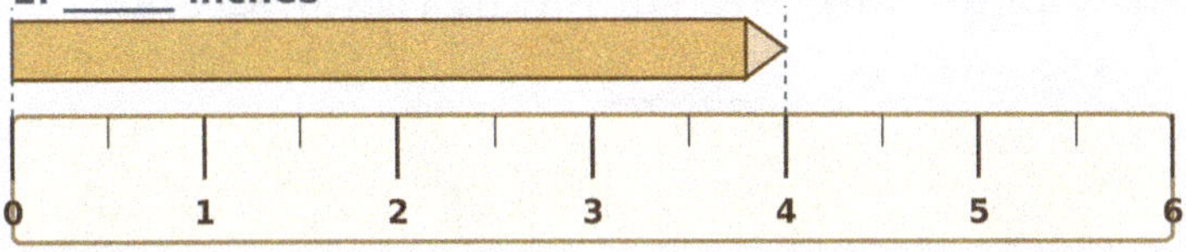

3. Measure a real pencil with an inch ruler. How long is it? ______ inches

REMEMBER

Write **in.** after a number of inches, with a period, so it is not read as the word “in”.

2 Measuring in Centimeters

Standard: 2.MD.A.2

NEW WORDS

centimeter — a metric unit of length, written cm

LET'S LEARN

A centimeter is smaller than an inch, so the same object measures a bigger number of centimeters.

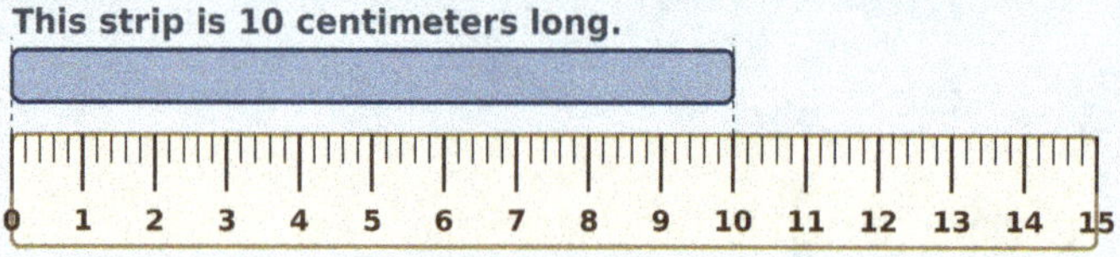

About 2 and a half centimeters fit inside one inch.

WORKED EXAMPLE

Step 1. Measure the ribbon in centimeters.

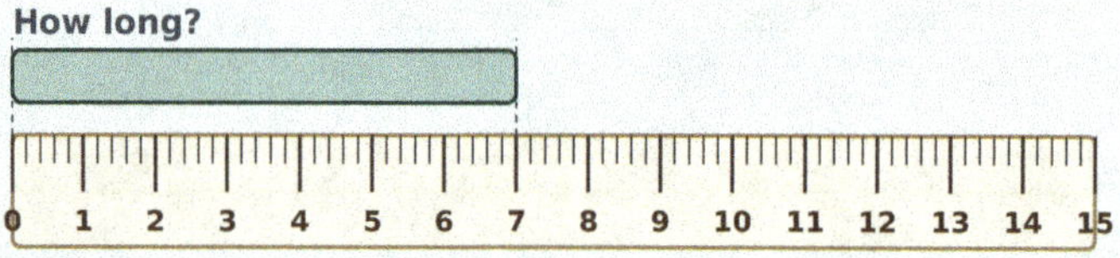

Step 2. The end reaches 7.

Answer: 7 cm.

MATH TALK

If a stick is 5 inches long, will the number of centimeters be more or fewer than 5? Why?

TRY IT TOGETHER

Write the length in centimeters.

1. _____ cm

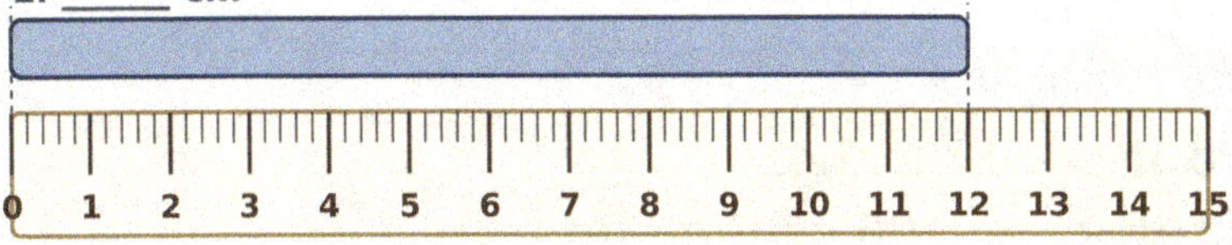

2. _____ cm

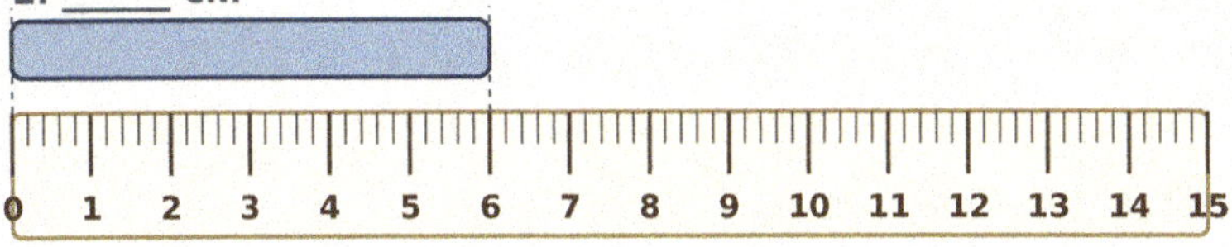

ON YOUR OWN

Write the length.

1. _____ cm

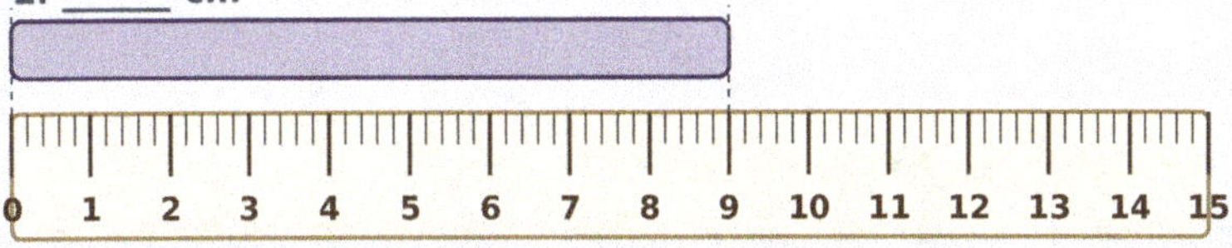

2. _____ cm

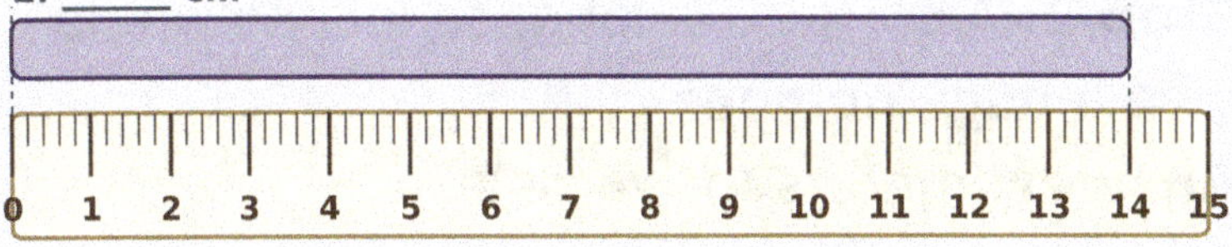

3. A pen is 6 inches long. Would it measure more or fewer than 6 centimeters?

____

3 Choosing a Unit

Standard: 2.MD.A.3

NEW WORDS

foot — 12 inches
meter — 100 centimeters
yard — 3 feet

LET'S LEARN

Small things use small units. Large things use large units.

WHICH UNIT?

inches — a crayon, a book
feet or yards — a desk, a hallway
centimeters — a paper clip, a finger
meters — a classroom, a playground

WORKED EXAMPLE

Step 1. Which unit suits a school corridor?
Step 2. A corridor is long, so inches would give a huge number.
Answer: meters (or yards).

MATH TALK

Why would measuring a playground in centimeters be a poor choice?

TRY IT TOGETHER

Write the best unit for each.

1. a pencil ______________
2. a classroom door ______________
3. a football field ______________

ON YOUR OWN

Choose inches, feet, centimeters, or meters.

1. a paper clip ______________
2. a bed ______________
3. a swimming pool ______________
4. a shoe ______________

5 **1 foot = ☐ inches**

6 **1 meter = ☐ cm**

4 Estimating Lengths

Standard: 2.MD.A.3

NEW WORDS

estimate — a sensible guess before measuring

LET'S LEARN

Use something you know well as a benchmark.

BENCHMARKS

Your finger is about **1 inch** wide.
A large paper clip is about **1 inch** long.
A doorway is about **1 meter** wide.
A big step is about **1 yard**.

WORKED EXAMPLE

Step 1. Estimate the length of a book.

Step 2. It looks about as long as 5 paper clips.

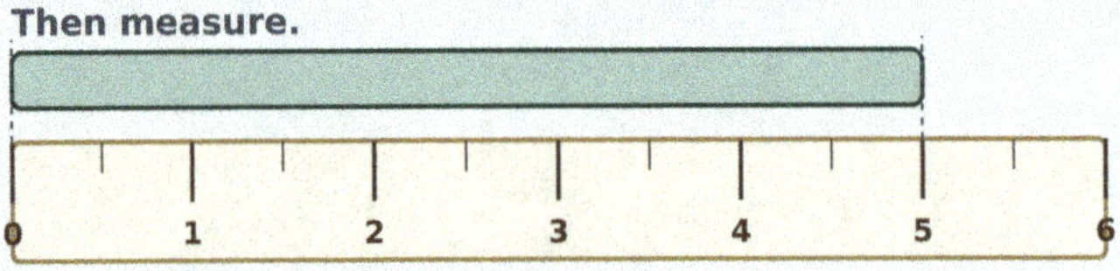

Answer: Estimate 5 in., actual 5 in.

MATH TALK

Why is a good estimate useful even when you plan to measure?

TRY IT TOGETHER

Circle the better estimate.

1. **a fork: 7 in. or 7 ft**
2. **a car: 4 m or 4 cm**

ON YOUR OWN

Circle the better estimate.

1. **a shoe: 9 in. or 9 ft**
2. **a room: 5 m or 5 cm**
3. **a crayon: 8 cm or 8 m**
4. **a bus: 10 m or 10 in.**

5. Estimate the width of your desk, then measure it.

Estimate	Measurement

5 Comparing Lengths

Standard: 2.MD.B.5

NEW WORDS

difference in length — how much longer one thing is

LET'S LEARN

To find how much longer, subtract the shorter length from the longer one.

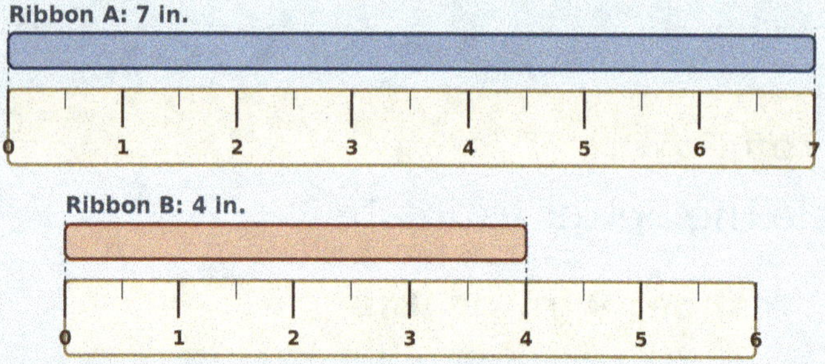

$$7 - 4 = 3$$

Ribbon A is 3 inches longer.

WORKED EXAMPLE

Step 1. A rope is 32 cm. Another is 18 cm.

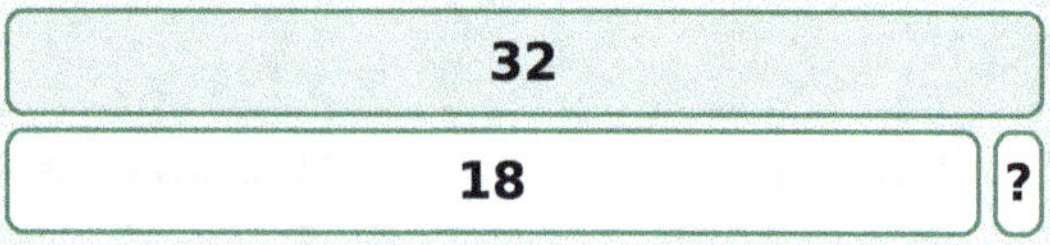

Step 2. Subtract to find the difference.

$$32 - 18 = 14$$

Answer: 14 cm longer.

MATH TALK

If two lengths are joined end to end, do you add or subtract?

TRY IT TOGETHER

1. A stick is 25 cm and another is 40 cm. How much longer is the second stick?

☐ – ☐ = ☐

ON YOUR OWN

2. A pencil is 15 cm and a pen is 14 cm. How much longer is the pencil? ______
3. Two ribbons of 26 in. and 35 in. are tied together. How long is the new ribbon? ______
4. A path is 60 m. Ana walks 45 m. How much further must she walk? ______
5. A table is 90 cm wide. A rug is 30 cm wider. How wide is the rug? ______

Math in My World: The Class Race

Problem solving with Unit 9 skills

Class 2B measured how far each paper airplane flew.

THE INFORMATION

Distances flown

Ana: **18 feet**

Ben: **25 feet**

Cara: **12 feet**

Dan: **25 feet**

YOUR TURN

1. Who flew the farthest? __________
2. How much farther did Ben fly than Cara? ______
3. Two children tied. Who were they? __________
4. What is the total distance for Ana and Cara together?

   ☐ + ☐ = ☐

5. Would feet or inches be the better unit for these distances? Why?

MATH TALK

Ben and Dan both flew 25 feet. How would you show that with a symbol?

Unit 9 Review

Review of Lessons 9.1–9.5

A Part A · Measure

1. ______ inches

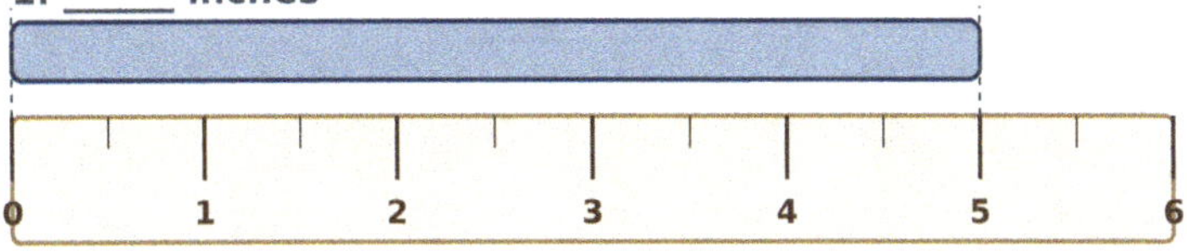

2. ______ cm

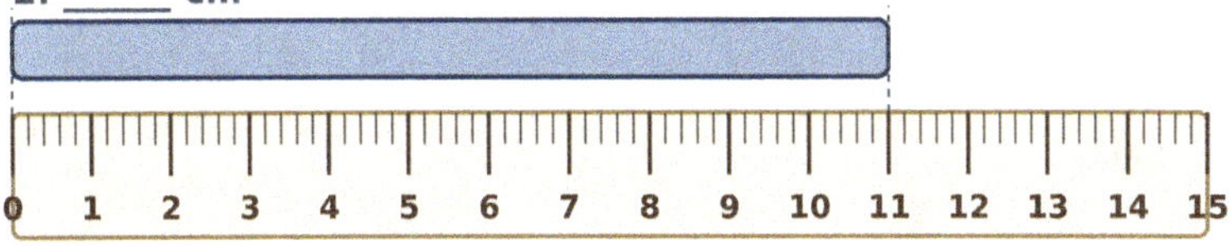

B Part B · Choose the unit

3. a book ______________

4. a garden ______________

C Part C · Estimate

5 **a door: 2 m or 2 cm**

6 **a spoon: 6 in. or 6 ft**

D Part D · Compare

7. A rope is 48 cm. Another is 29 cm. How much longer is the first rope? ______

8. Two strips of 34 in. and 27 in. are joined. How long is the new strip? ______

E Part E · Units

9 **1 foot = ☐ inches**

10 **1 meter = ☐ cm**

CHALLENGE

A worm is 8 cm long. A pencil is 3 cm longer than the worm. A ruler is twice as long as the pencil. How long is the ruler?

UNIT
10

Working with Data

YOUR LEARNING GOALS

In this unit you will learn to...

- Show measurement data on a line plot.
- Read a picture graph with a key.
- Read and draw a bar graph.
- Compare the bars to answer questions.
- Collect data and show it your own way.

STANDARDS COVERED

2.MD.D.9 Generate measurement data and show the measurements on a line plot.
2.MD.D.10 Draw a picture graph and a bar graph to represent a data set with up to four categories.
2.OA.A.1 Solve put-together, take-apart, and compare problems.

1 Line Plots

Standard: 2.MD.D.9

NEW WORDS

line plot — a number line with a mark for each measurement

data — the measurements or facts you collect

LET'S LEARN

A **line plot** puts one mark above the number for every measurement.

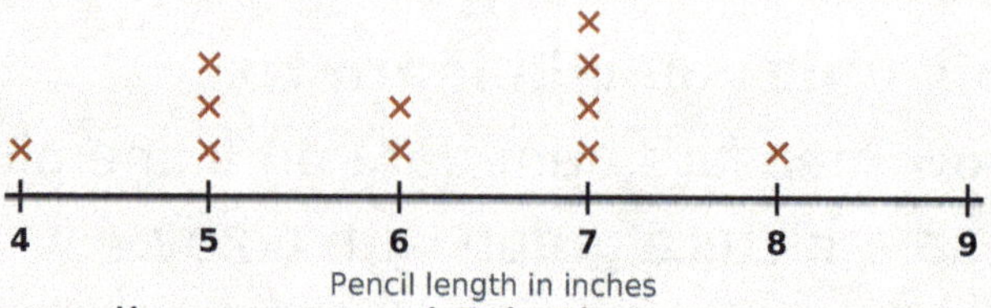

Four pencils measured 7 inches, so there are four marks above the 7.

WORKED EXAMPLE

Step 1. How many pencils were measured in all?

Step 2. Count every mark: 1 + 3 + 2 + 4 + 1.

Answer: 11 pencils.

MATH TALK

What does a tall stack of marks tell you about that length?

TRY IT TOGETHER

Use the line plot to answer.

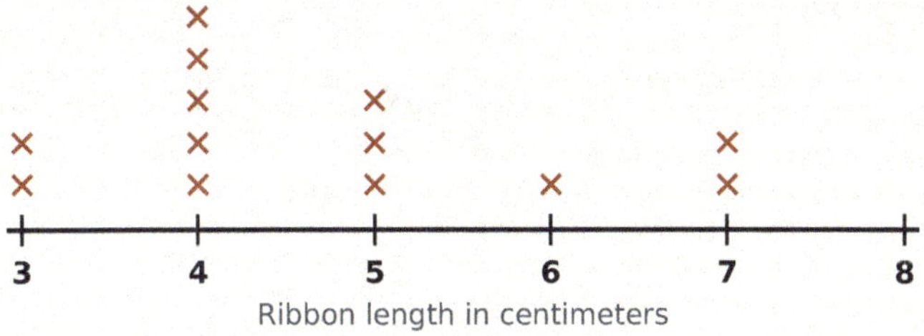

1 **Most common length:** ☐

2 **How many ribbons in all:** ☐

ON YOUR OWN

Use the line plot above.

1. How many ribbons were 5 cm long? ______
2. How many ribbons were longer than 5 cm? ______
3. Which length had the fewest ribbons? ______
4. How many more ribbons were 4 cm than 6 cm? ______
5. Measure five pencils in your class. Draw a line plot of your results.

2 Picture Graphs

Standard: 2.MD.D.10

NEW WORDS

key — a note that says how much one symbol stands for

LET'S LEARN

In a picture graph, each symbol can stand for more than one.

Monday	★ ★ ★
Tuesday	★ ★ ★ ★ ★
Wednesday	★ ★

Key: each symbol = 2

Each star stands for 2 books, so 5 stars means 10 books.

WORKED EXAMPLE

Step 1. Tuesday has 5 stars.

Step 2. The key says each star is 2 books.

$$5 \times 2 = 10$$

Answer: 10 books on Tuesday.

MATH TALK

Why might a graph maker choose a key of 2 instead of 1?

TRY IT TOGETHER

Use the graph above.

1 **Books on Monday:** ☐

2 **Books on Wednesday:** ☐

3 **Books in all:** ☐

ON YOUR OWN

Use this graph.

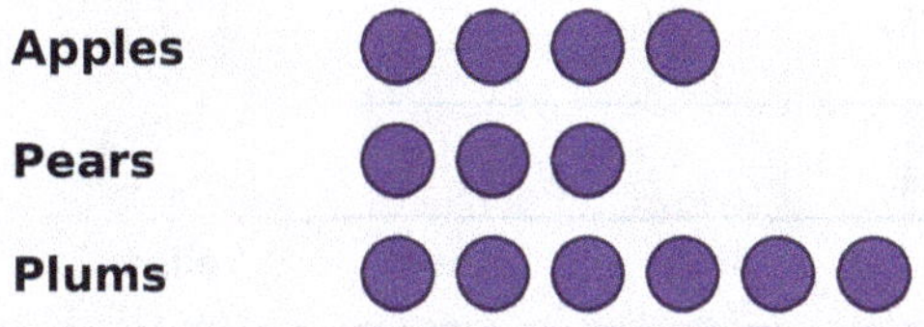

Key: each symbol = 2

1. How many plums? ______
2. How many more plums than pears? ______
3. How many pieces of fruit in all? ______
4. If the key changed to 1 fruit per symbol, how many symbols would the apple row need? ______

3 Bar Graphs

Standard: 2.MD.D.10

NEW WORDS

bar graph — a graph that shows amounts as bars

scale — the numbers up the side of a graph

LET'S LEARN

The height of each bar shows how many. Read across to the scale.

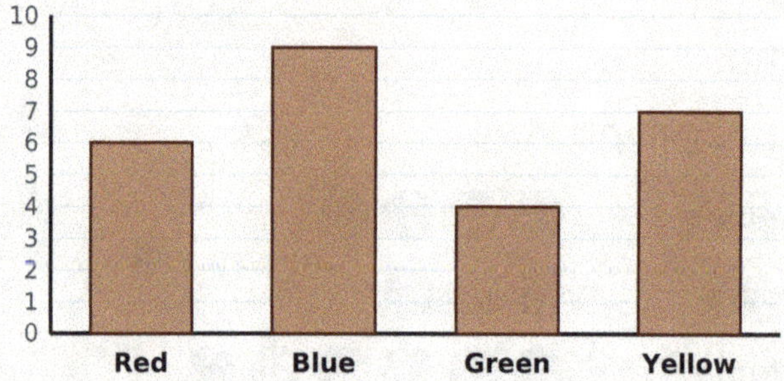

The blue bar reaches 9, so 9 children chose blue.

WORKED EXAMPLE

Step 1. Which color was chosen least?

Step 2. Find the shortest bar.

Answer: Green, with 4 children.

MATH TALK

Why must every bar be measured from the same baseline?

TRY IT TOGETHER

Use the graph above.

1 **Children who chose red:** ☐

2 **Most popular color:** ☐

3 **Children in all:** ☐

ON YOUR OWN

Use this graph.

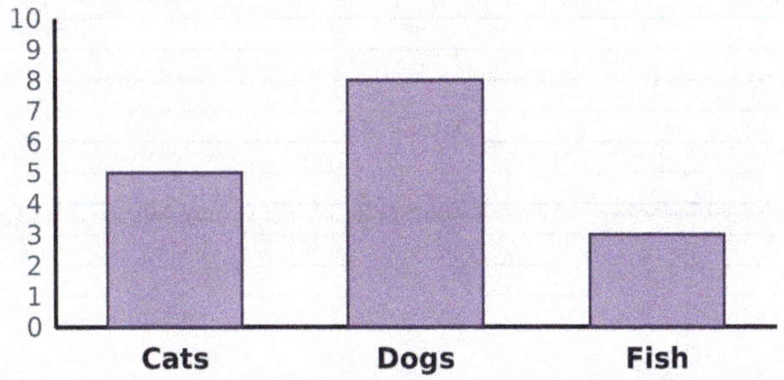

1. How many dogs? ______
2. How many more dogs than fish? ______
3. How many pets in all? ______
4. Two more cats join. How many cats now?

______

4 Comparing on Graphs

Standard: 2.OA.A.1

NEW WORDS

compare — to find how much more or less

LET'S LEARN

Graph questions are addition and subtraction problems in disguise.

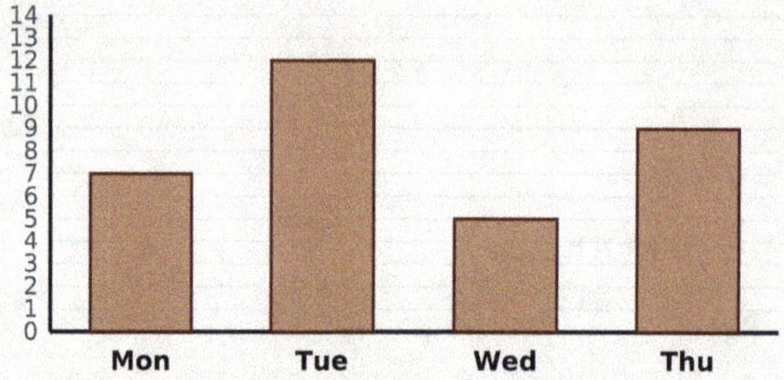

$$12 - 5 = 7$$

Tuesday had 7 more than Wednesday.

WORKED EXAMPLE

Step 1. How many on Monday and Thursday together?

$$7 + 9 = 16$$

Step 2. Add the two bar heights.

Answer: 16.

MATH TALK

Which words in a question tell you to subtract?

TRY IT TOGETHER

Use the graph above.

1. How many more on Thursday than Monday? ______
2. How many in all for the four days? ______

ON YOUR OWN

Use this graph.

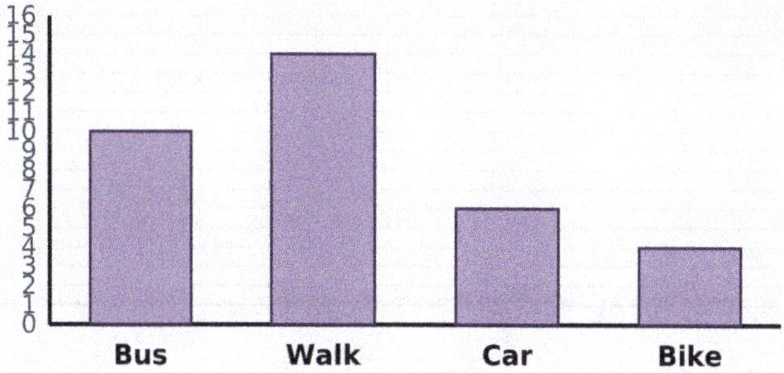

1. How many children walk? ______
2. How many more walk than come by car? ______
3. How many come by bus or bike? ______
4. How many children are in the survey altogether? ______
5. Which two ways together equal the number who walk? __________

5 Collecting Data

Standard: 2.MD.D.10

NEW WORDS

survey — asking the same question of many people
category — one of the groups you sort answers into

LET'S LEARN

First collect with tallies, then show the data in a graph.

Apple	𝍸 ||
Banana	𝍸 𝍸 |
Orange	||||

Tallies are quick to write while you are counting.

WORKED EXAMPLE

Step 1. Read the tallies: 7, 11, and 4.
Step 2. Draw bars of those heights.

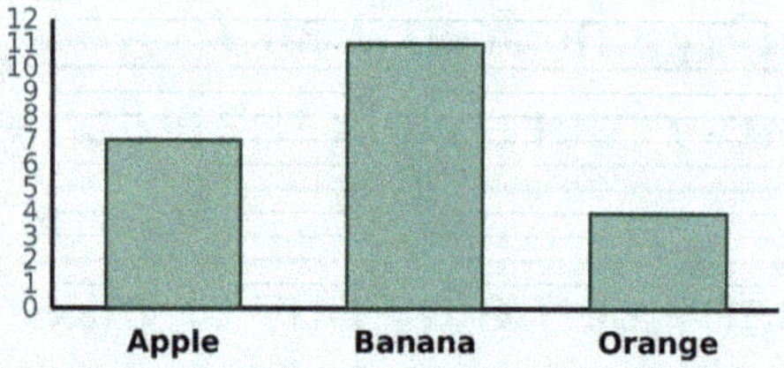

Answer: The bar graph shows the same data.

MATH TALK

Why are tallies easier than numerals while you are collecting?

TRY IT TOGETHER

Write the number for each tally.

Red  ☐

Blue  ☐

ON YOUR OWN

1. Survey ten classmates about a favorite sport. Record your results.

Sport	Tally	Number

2. Which category had the most? ____________

3. How many people did you ask in all?

   ____

REMEMBER

Every graph needs a title and clear labels, or a reader cannot tell what it is about.

★

Math in My World: The Weather Chart

Problem solving with Unit 10 skills

Class 2B recorded the weather for four weeks.

THE GRAPH

Days of each kind of weather

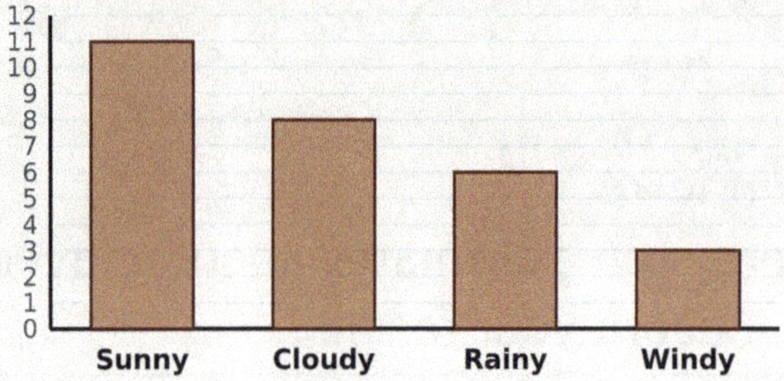

YOUR TURN

1. How many sunny days? ______
2. How many more sunny days than rainy days?

______

3. How many cloudy and windy days together?

______

4. How many days were recorded in all? ______
5. Which two kinds of weather together equal the number of sunny days? __________

MATH TALK

Would a picture graph with a key of 2 show this data well? What problem would the windy row cause?

Unit 10 Review

Review of Lessons 10.1–10.5

A Part A · Line plot

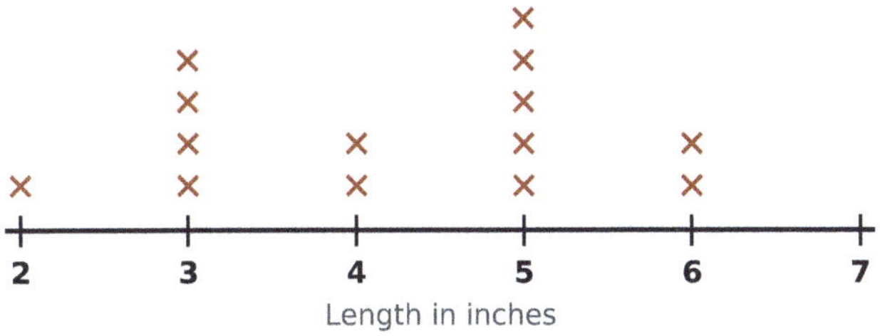

1. Most common length: ______

2. How many objects in all? ______

B Part B · Picture graph

3. How many dogs? ______

4. How many more dogs than cats? ______

C Part C · Bar graph

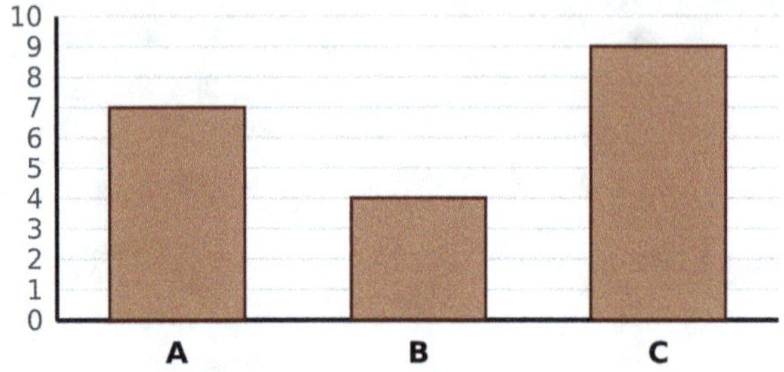

5. Which is greatest? ______

6. How many more in C than B? ______

7. How many in all? ______

D Part D · Tallies

Votes 

CHALLENGE

A bar graph has four bars. Two of them add to 15, and all four add to 28. If the other two bars are equal, how tall is each one?

UNIT

11

Shapes and Arrays

YOUR LEARNING GOALS

In this unit you will learn to...

- Name flat shapes by their sides and angles.
- Name solid shapes by their faces and edges.
- Draw shapes with given attributes.
- Split a rectangle into rows and columns of squares.
- Use repeated addition to find the total in an array.

STANDARDS COVERED

2.G.A.1 Recognize and draw shapes having specified attributes, such as a given number of angles or faces.
2.G.A.2 Partition a rectangle into rows and columns of same-size squares and count to find the total.
2.OA.C.4 Use addition to find the total number of objects in a rectangular array.

1 Sides and Angles

Standard: 2.G.A.1

NEW WORDS

angle — the corner where two sides meet
side — one straight edge of a shape

LET'S LEARN

A shape is named by how many sides and angles it has, not by its size or color.

3 sides **4 sides** **5 sides** **6 sides**

A shape always has the same number of angles as sides.

WORKED EXAMPLE

Step 1. A shape has 4 sides and 4 angles.

Step 2. If all 4 sides are equal, it is a square. If only opposite sides are equal, it is a rectangle.

square **rectangle**

Answer: Both are four-sided shapes.

MATH TALK

Is every square also a rectangle? Explain what you think.

TRY IT TOGETHER

Write how many sides and angles.

1 **triangle:** ☐ **sides,** ☐ **angles**

2 **hexagon:** ☐ **sides,** ☐ **angles**

ON YOUR OWN

Name the shape.

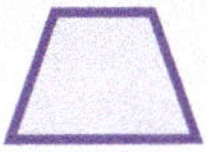 ____

 ____

 ____

Answer.

1. I have 6 sides. What am I? __________
2. I have 3 angles. What am I? __________
3. I have no sides and no angles. What am I?

____

4. Draw a shape with exactly 5 angles.

2 Faces and Edges

Standard: 2.G.A.1

NEW WORDS

face — a flat surface of a solid
edge — where two faces meet
vertex — a corner point of a solid

LET'S LEARN

Solid shapes take up space. We count their **faces**, **edges**, and **vertices**.

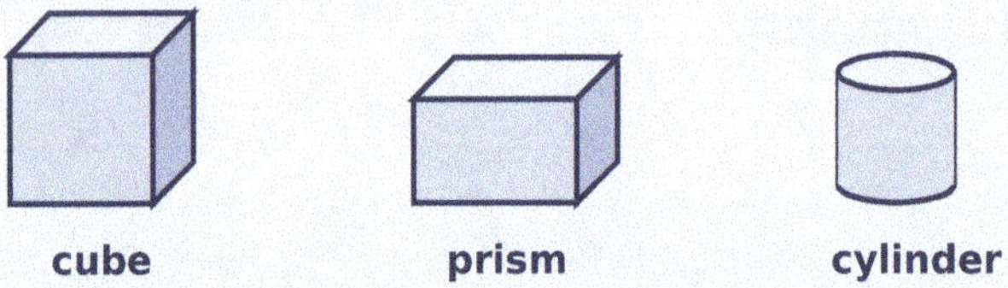

A cube has 6 faces, 12 edges, and 8 vertices. Every face is a square.

WORKED EXAMPLE

Step 1. How many faces does a rectangular prism have?

Step 2. Top and bottom, front and back, two sides.

Answer: 6 faces, and each one is a rectangle.

MATH TALK

Which solids can roll? What do their surfaces have in common?

TRY IT TOGETHER

Name each solid.

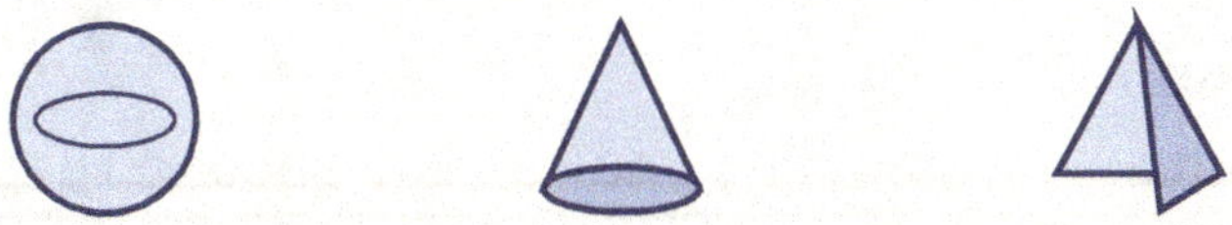

_____ _____ _____

ON YOUR OWN

Answer.

1 **A cube has ☐ faces.**

2 **A cube has ☐ edges.**

3 **A cube has ☐ vertices.**

4. Which solid is like a can of soup?

____

5. Which solid is like a party hat? __________

6. Which solid has no flat faces at all?

____

REMEMBER

A face is flat, an edge is a line, and a vertex is a point. Count them in that order.

3 Drawing Shapes

Standard: 2.G.A.1

NEW WORDS

attribute — something you can say about a shape

LET'S LEARN

To draw a shape from a description, count the sides as you draw them.

5 sides, 5 angles

Start with one side, then keep going until the shape closes.

WORKED EXAMPLE

Step 1. Draw a shape with 4 sides where only opposite sides are equal.

rectangle

Step 2. Two long sides and two short sides.
Answer: a rectangle.

MATH TALK

Can you draw a shape with 4 sides that is not a rectangle? What would it look like?

TRY IT TOGETHER

Draw each shape.

1. a shape with 3 angles   2. a shape with 6 sides

ON YOUR OWN

Draw a shape for each description.

1. 4 equal sides   2. exactly 5 angles   3. no straight sides

4. Name a real object shaped like a rectangle.

____

4 Rows and Columns of Squares

Standard: 2.G.A.2

NEW WORDS

partition — to split into equal parts
row — a line across
column — a line down

LET'S LEARN

A rectangle can be split into equal squares. Count the rows and the columns.

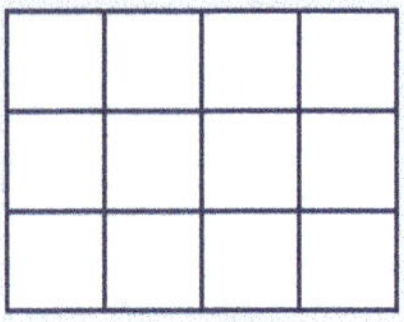

3 rows, 4 columns

$$4 + 4 + 4 = 12$$

There are 12 squares in all.

WORKED EXAMPLE

Step 1. This rectangle has 2 rows and 5 columns.

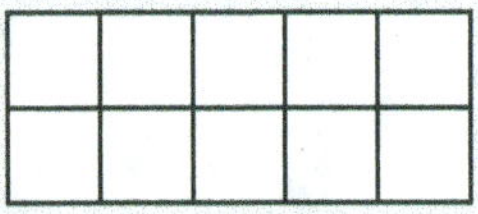

2 rows, 5 columns

Step 2. Add the rows: 5 + 5.

Answer: 10 squares.

MATH TALK

Would 4 rows of 3 give the same number of squares as 3 rows of 4? Explain.

TRY IT TOGETHER

Count the rows, the columns, and the squares.

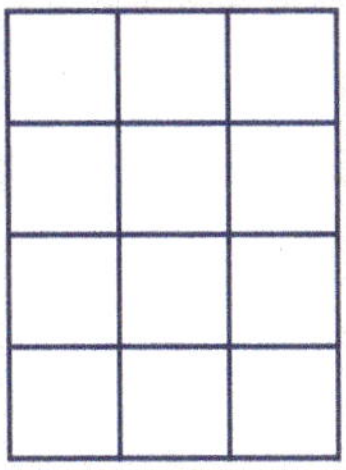

1 **rows:** **columns:** **squares:**

ON YOUR OWN

Write how many squares.

1. 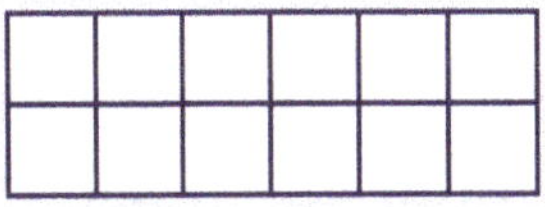

2. 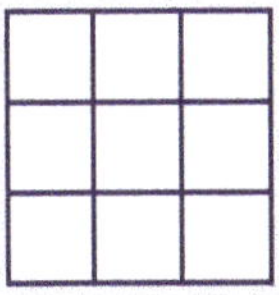

3. Draw a rectangle with 2 rows and 4 columns of squares. How many squares?

____

5 Arrays and Repeated Addition

Standard: 2.OA.C.4

NEW WORDS

array — objects in equal rows and columns

LET'S LEARN

Add one row at a time to find the total in an array.

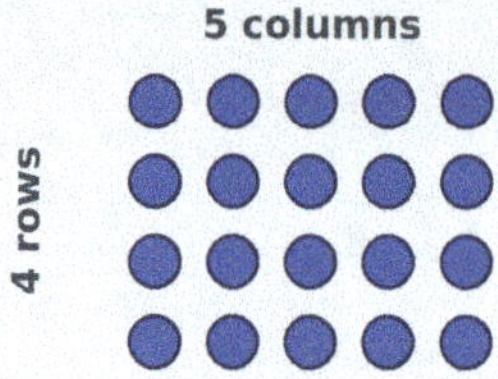

5 + 5 + 5 + 5 = 20

Four rows of five is twenty.

WORKED EXAMPLE

Step 1. Count the rows: 3. Count each row: 6.

Step 2. Add: 6 + 6 + 6.

Answer: 18.

MATH TALK

How is repeated addition different from ordinary addition?

TRY IT TOGETHER

Write the addition sentence and the total.

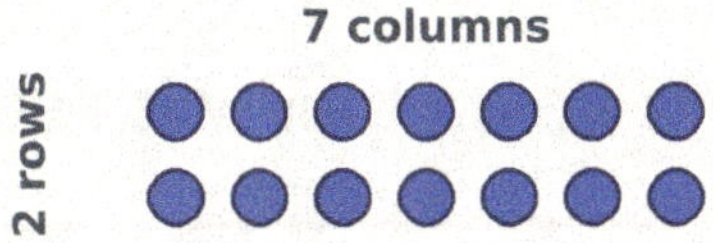

☐ + ☐ = ☐

ON YOUR OWN

Write the total.

1.  ☐

2. 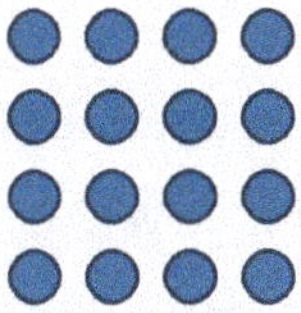 ☐

3. A tray has 5 rows of 4 muffins. How many muffins? ______

4. Draw an array with 3 rows of 7. Write the addition sentence.

Math in My World: The Tile Floor

Problem solving with Unit 11 skills

A kitchen floor is being tiled with square tiles.

THE PICTURE

The floor

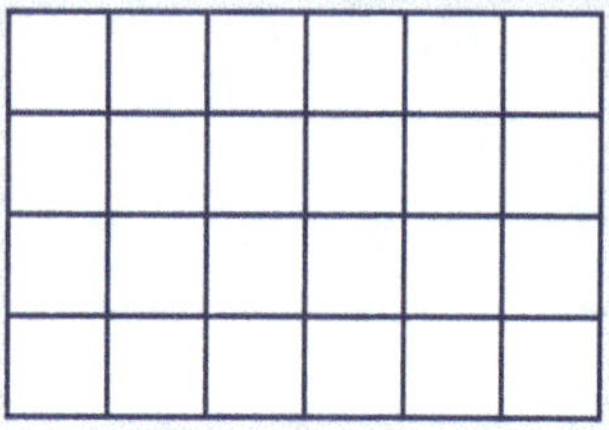

4 rows, 6 columns

YOUR TURN

1. How many tiles are in one row? ______
2. How many rows are there? ______
3. Write an addition sentence for the whole floor.

   ....

4. How many tiles in all? ______
5. What shape is each tile? How many angles does it have?

   ....

MATH TALK

If the floor were turned so it had 6 rows and 4 columns, would more tiles be needed? Explain.

Unit 11 Review

Review of Lessons 11.1–11.5

A Part A · Name the shape

______ ______ ______

B Part B · Solids

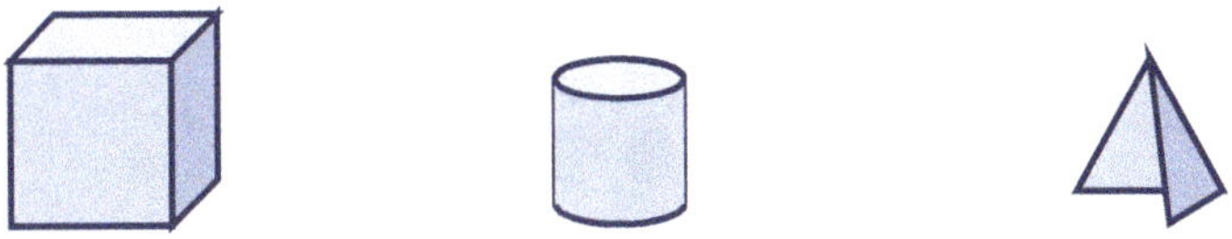

______ ______ ______

1. A cube has ______ faces and ______ edges.

C Part C · Draw

2. Draw a shape with 5 sides.

D Part D · Rows and columns

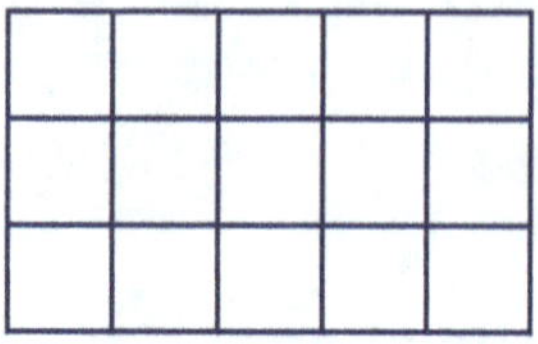

3. rows: ______ columns: ______ squares: ______

E Part E · Arrays

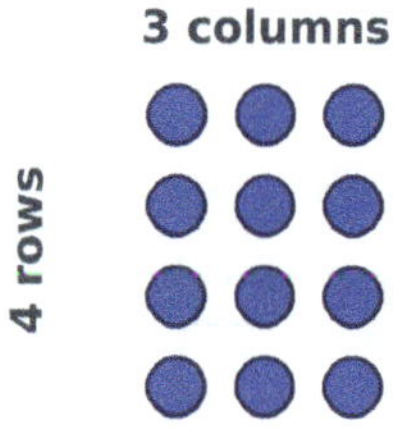

☐ + ☐ + ☐ + ☐ = ☐

CHALLENGE

A rectangle is split into 20 equal squares with 4 rows. How many columns does it have? Draw it.

UNIT

12

Equal Shares

YOUR LEARNING GOALS

In this unit you will learn to...

- Split a shape into two equal shares.
- Split a shape into three equal shares.
- Split a shape into four equal shares.
- See that equal shares of the same whole are the same size.
- Use equal shares to solve everyday problems.

STANDARDS COVERED

2.G.A.3 Partition circles and rectangles into two, three, or four equal shares; describe the shares as halves, thirds, and fourths; recognize that equal shares of identical wholes need not have the same shape.

1 Halves

Standard: 2.G.A.3

NEW WORDS

half — one of two equal shares
equal shares — parts that are the same size

LET'S LEARN

Two **equal** shares are called halves. Both parts must be the same size.

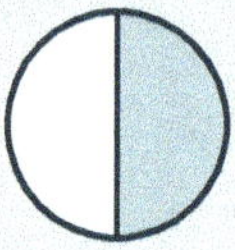

one half

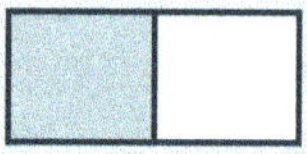

one half

Two halves make one whole.

WORKED EXAMPLE

Step 1. Split a sandwich between two children.

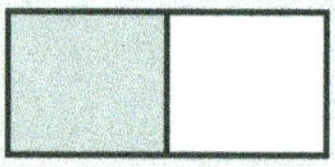

halves

Step 2. Each child gets one of two equal shares.
Answer: Each child gets one half.

MATH TALK

If a shape is cut into two parts of different sizes, are they halves? Why not?

TRY IT TOGETHER

Circle the shapes that show halves.

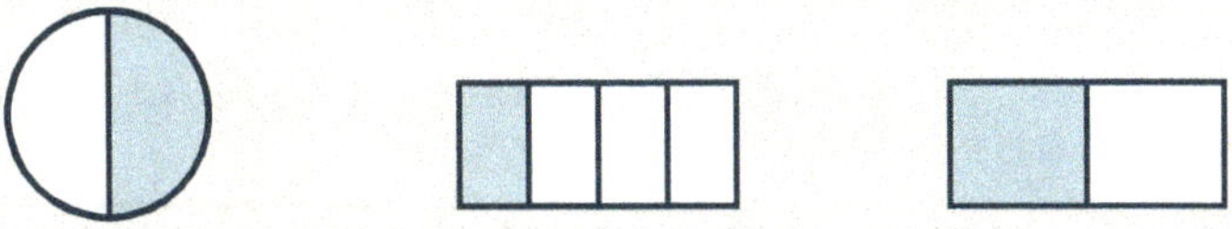

ON YOUR OWN

Shade one half of each shape.

1. How many equal shares are in a half?

   ______

2. Two halves make one ______
3. Is half a large pizza the same size as half a small pizza? Explain.

2 Thirds

Standard: 2.G.A.3

NEW WORDS

third — one of three equal shares

LET'S LEARN

Three equal shares are called **thirds**.

one third

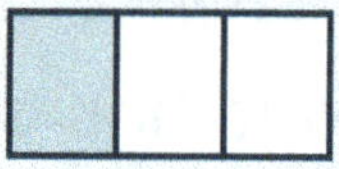

one third

Three thirds make one whole.

WORKED EXAMPLE

Step 1. Three friends share a pie equally.

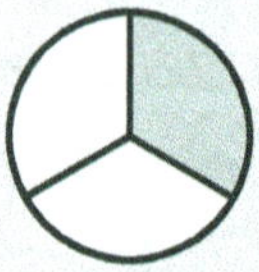

thirds

Step 2. The pie is cut into three equal shares.

Answer: Each friend gets one third.

MATH TALK

Which is bigger, one half or one third of the same shape? Why?

TRY IT TOGETHER

Write **halves**, **thirds**, or **not equal**.

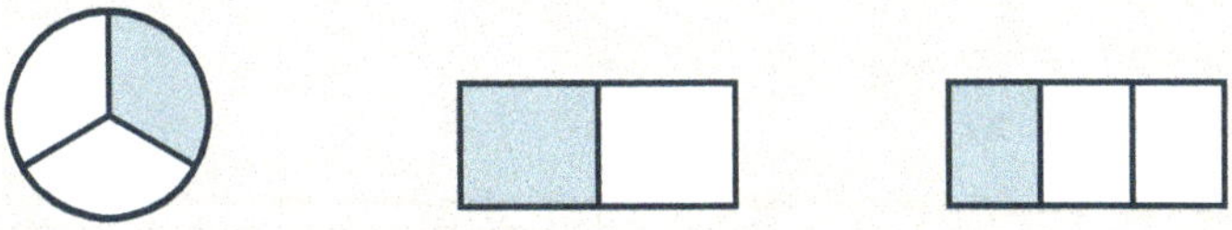

ON YOUR OWN

Shade one third of each shape.

1. How many equal shares are in thirds?

______

2. Three thirds make one ______
3. A ribbon is cut into thirds. How many pieces are there? ______
4. Draw a rectangle split into thirds a different way from the picture above.

3 Fourths

Standard: 2.G.A.3

NEW WORDS

fourth — one of four equal shares
quarter — another name for a fourth

LET'S LEARN

Four equal shares are called **fourths** or **quarters**.

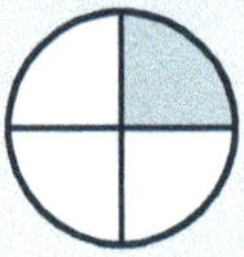

one fourth

one fourth

Four fourths make one whole. Two fourths are the same as one half.

WORKED EXAMPLE

Step 1. A cake is cut for four children.

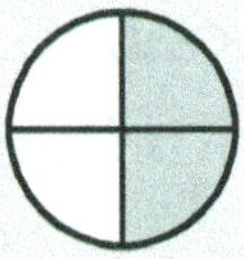

two fourths shaded

Step 2. Two of the four shares are taken.

Answer: Two fourths, which is the same as one half.

MATH TALK

Why do two fourths make one half?

TRY IT TOGETHER

Write how many equal shares.

ON YOUR OWN

Shade one fourth of each shape.

1. How many fourths make one whole? ______
2. How many fourths make one half? ______
3. Which is bigger, one third or one fourth? __________
4. Four children share a pizza equally. Each gets one __________

REMEMBER

More shares means smaller pieces: a fourth is smaller than a third, which is smaller than a half.

4 Same Whole, Different Shapes

Standard: 2.G.A.3

NEW WORDS

identical wholes — wholes that are exactly the same size

LET'S LEARN

Equal shares of the same whole are the same **size**, but they need not be the same **shape**.

cut down **cut across**

Both squares show one half, cut in different directions.

WORKED EXAMPLE

Step 1. Two identical sandwiches are cut in half.
Step 2. One is cut down the middle, the other across.
Answer: The two halves look different but hold the same amount of food.

MATH TALK

Two children each eat half of their own identical sandwich. Do they eat the same amount?

TRY IT TOGETHER

Do these two shapes show equal shares of the same size? Write **yes** or **no**.

ON YOUR OWN

1. Draw two identical rectangles. Split one into halves down the middle and the other into halves across the middle.

2. Are the two halves the same size? ______
3. Are they the same shape? ______
4. Ana ate half of a small cake. Ben ate half of a large cake. Did they eat the same amount? Explain.

5 Equal Shares Every Day

Standard: 2.G.A.3

NEW WORDS

share equally — to give everyone the same amount

LET'S LEARN

Sharing fairly means everyone gets an equal share.

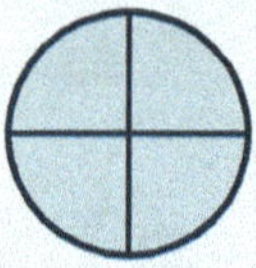

one pizza, four equal shares

Four children, four equal shares: each gets one fourth.

WORKED EXAMPLE

Step 1. Three children share a chocolate bar.

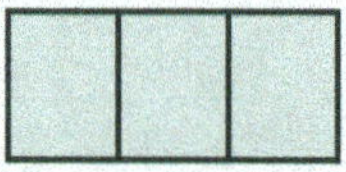

thirds

Step 2. Cut it into three equal pieces.

Answer: Each child gets one third.

MATH TALK

Why must the shares be equal for the sharing to be fair?

TRY IT TOGETHER

Write how much each person gets.

1. 2 children share 1 apple. Each gets one ______
2. 4 children share 1 cake. Each gets one ______
3. 3 children share 1 pie. Each gets one ______

ON YOUR OWN

Answer.

1. A pizza is cut into fourths. Two are eaten. How many are left? ______
2. Is two fourths the same as one half? ______
3. A sandwich is cut into thirds. One third is eaten. How many thirds are left? ______
4. Draw a rectangle and show how three children could share it equally.

Math in My World: The Birthday Party

Problem solving with Unit 12 skills

There are treats to share at a birthday party.

THE PICTURE

The treats

YOUR TURN

1. The pizza is shared by 4 children. Each gets one ______
2. The cake is shared by 3 children. Each gets one ______
3. The melon is shared by 2 children. Each gets one ______
4. Which treat gives the largest share to each child? __________
5. Two children eat their shares of the pizza. How much pizza is left? __________
6. Shade one fourth of the pizza in the picture above.

MATH TALK

All three treats were shared equally, yet the shares are different sizes. Explain why.

Unit 12 Review

Review of Lessons 12.1–12.5

A Part A · Name the shares

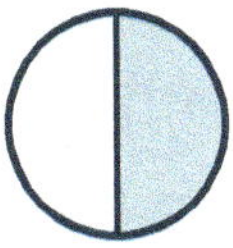

______

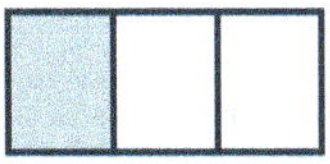

______

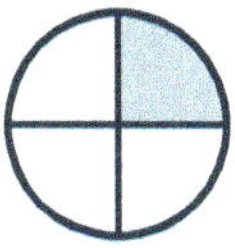

______

B Part B · Shade

Shade one third of the first shape and one fourth of the second.

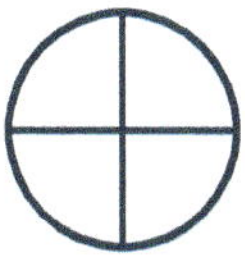

C Part C · Questions

1. How many fourths make one whole? ______

2. How many thirds make one whole? ______

D Part D · Compare shares

3. Which is bigger, one half or one fourth?

____

4. Two fourths is the same as one ______

E Part E · Same whole

5. Two identical squares are each cut in half, one down and one across. Are the halves the same size? Explain.

F Part F · Real life

6. Three friends share one sandwich equally. Each gets one ______

CHALLENGE

Draw a rectangle. Show two different ways to split it into fourths. Explain why both are correct.

Glossary

Every word in this list appears in a **New words** box somewhere in the book.

a.m.	the hours from midnight to noon
angle	the corner where two sides of a shape meet
array	objects arranged in equal rows and columns
bar graph	a graph that shows amounts as bars
category	one of the groups that data is sorted into
cent	the smallest unit of money, written ¢
centimeter	a metric unit of length, written cm
column	a line of objects going down
compare	to decide which number is greater, less, or equal
data	the facts or measurements you collect
difference	the answer when you subtract
digit	one of the symbols 0, 1, 2, 3, 4, 5, 6, 7, 8, or 9
dollar	one hundred cents, written $1
edge	the line where two faces of a solid meet
equal shares	parts of a whole that are exactly the same size
estimate	a sensible guess made before measuring or calculating
even	a number that can be split into two equal groups

Glossary *continued*

expanded form	a number written as the sum of its place values, such as 300 + 40 + 7
face	a flat surface of a solid shape
fact family	the four facts made by three related numbers
foot	a unit of length equal to 12 inches
fourth	one of four equal shares; also called a quarter
half	one of two equal shares
hundred	ten groups of ten
inch	a customary unit of length, written in.
key	a note on a graph saying how much one symbol stands for
line plot	a number line with a mark for each measurement
meter	a metric unit of length equal to 100 centimeters
minute hand	the long hand on a clock
nickel	a coin worth 5 cents
odd	a number that has one left over when split into two groups
open number line	a number line with only the numbers you need marked on it
p.m.	the hours from noon to midnight
partition	to split a shape into equal parts

Glossary *continued*

penny	a coin worth 1 cent
place value	what a digit is worth because of where it sits
quarter	a coin worth 25 cents; also one of four equal shares
quarter past	fifteen minutes after the hour
quarter to	fifteen minutes before the next hour
regroup	to trade ten of one place for one of the next place
repeated addition	adding the same number again and again
row	a line of objects going across
ruler	a measuring tool marked in equal units
scale	the numbers up the side of a graph
skip-count	to count in equal jumps, such as 5, 10, 15
sum	the answer when you add
survey	asking the same question of many people
third	one of three equal shares
vertex	a corner point of a solid shape
yard	a unit of length equal to 3 feet

Answer Key

Answers are listed unit by unit. Where a question has more than one correct answer, an example is given.

Unit 1 · Lesson 1 — Tens and Ones

Try it together: 35; 62

On your own: 1. 2 tens 7 ones 2. 8 tens 4 ones 3. 6 tens 0 ones 4. 59 5. 90 6. 17 7. 74

Math talk: The 5 is worth 50 (five tens). The 8 is worth 8 ones.

Unit 1 · Lesson 2 — Counting Patterns

Try it together: 15, 20, 25, 30, 35; 46, 47, 48, 49, 50

On your own: fives 25, 30, 35, 40, 45; tens 18, 28, 38, 48, 58; back 73, 72, 71, 70, 69; 1. 66 2. 81

Math talk: The ones digit stays the same; only the tens digit grows.

Unit 1 · Lesson 3 — Comparing Numbers

Try it together: 1. < 2. = 3. > 4. <

On your own: 1. > 2. < 3. = 4. > 5. < 6. >; order 27, 70, 72

Math talk: No. A number with more tens is always greater, because each ten is worth 10 ones.

Unit 1 · Lesson 4 — Even and Odd

Try it together: 1. even 2. odd 3. even 4. odd

On your own: even are 12, 20, 48, 64; 9. 8 + 8 10. 9 + 9

Math talk: An even number splits into two equal groups, and those two equal groups are exactly what a double is.

Unit 1 · Lesson 5 — Rows and Columns

Try it together: 6 + 6 = 12

On your own: 1. 12 2. 10; an array of 4 rows of 3 with total 12 (3 + 3 + 3 + 3 = 12)

Math talk: Yes — the total is 15 either way. Turning the array on its side does not change how many objects there are.

Unit 1 · Math in My World: The School Garden

1. 18 2. 6 + 6 + 6 = 18 3. 47 4. 4 tens 7 ones 5. even 6. the packet (47 > 45)

Unit 1 Review

Part A: 54; 1. 6 tens 8 ones 2. 39

Part B: 35, 40, 45, 50, 55; 16, 26, 36, 46, 56

Part C: 3. > 4. <

Part D: 5. even 6. odd 7. even 8. odd

Part E: 4 + 4 + 4 + 4 = 16

Challenge: greatest 63, least 36, and 63 > 36.

Answer Key *continued*

Unit 2 · Lesson 1 — Doubles and Near Doubles

Try it together: 1. 10, 11 2. 18, 17

On your own: 1. 8 2. 12 3. 16 4. 15 5. 11 6. 19 7. 4 + 4 (then add one)

Math talk: 8 + 8 = 16, then one more makes 17. (Or 9 + 9 = 18, then one less.)

Unit 2 · Lesson 2 — Make a Ten

Try it together: 1. 2, 12 2. 4, 14

On your own: 1. 13 2. 15 3. 13 4. 17 5. 14 6. 18; open number line 8 → 10 → 17

Math talk: Adding to 10 needs no regrouping — the answer is just ten and the leftover ones.

Unit 2 · Lesson 3 — Subtraction Facts

Try it together: 1. 6, 6 2. 7, 7

On your own: 1. 8 2. 8 3. 9 4. 8 5. 9 6. 7 7. 6 8. 9

Math talk: 13 − 6 asks what joins 6 to make 13, and that is the 7 from the addition fact.

Unit 2 · Lesson 4 — Fact Families

Try it together: 6 + 8 = 14, 8 + 6 = 14, 14 − 6 = 8, 14 − 8 = 6

On your own: 1. 8 2. 9 3. 8 4. 14 5. 13 6. 9

Math talk: Both start with the whole, because subtracting means taking one part away from the whole.

Unit 2 · Lesson 5 — Adding Three Numbers

Try it together: 1. 16 2. 19

On your own: 1. 17 2. 15 3. 15 4. 16 5. 18 books

Math talk: No. Addition can be done in any order, so the total never changes.

Unit 2 · Math in My World: The Sports Day

1. 8 2. 8 + 6 = 14 3. 3 4. 23 5. 14

Unit 2 Review

Part A: 1. 14 2. 17 3. 11 4. 16

Part B: 5. 8 6. 8 7. 7 8. 9

Part C: 5 + 8 = 13, 8 + 5 = 13, 13 − 5 = 8, 13 − 8 = 5

Part D: 9. 18 10. 16

Part E: 11. 8 12. 13

Part F: 13. 12 marbles

Challenge: e.g. 4 + 9 = 13 with 13 − 4 = 9, and 6 + 7 = 13 with 13 − 6 = 7.

Unit 3 · Lesson 1 — Adding Tens and Ones

Try it together: 1. 68 2. 87

On your own: 1. 36 2. 76 3. 87 4. 26 5. 86 6. 69 7. 87

Math talk: Tens and ones are separate places, so adding them separately never mixes them up.

Answer Key *continued*

Unit 3 · Lesson 2 — The Open Number Line

Try it together: 38, 48, 52; 28 + 24 = 52

On your own: 35 → 55 → 60 → 62, so 35 + 27 = 62; 1. 69 2. 89 3. 72 4. 82

Math talk: Landing on a ten makes the next jump easy, because adding to a ten needs no regrouping.

Unit 3 · Lesson 3 — Breaking Numbers Apart

Try it together: 40 + 6 and 30 + 3; 46 + 33 = 79

On your own: 1. 62 2. 83 3. 63 4. 94 5. 70 + 3

Math talk: The 12 ones become 1 ten and 2 ones, so the extra ten joins the tens: 70 + 10 + 2 = 82.

Unit 3 · Lesson 4 — Regrouping Ones

Try it together: 63; 63

On your own: 82; 83; 94; 1. 85 2. 82

Math talk: Add the ones digits first in your head. If they reach 10 or more, a regroup is coming.

Unit 3 · Lesson 5 — Adding Several Numbers

Try it together: 1. 68 2. 51

On your own: 1. 66 2. 55 3. 65 4. 60 5. 78 books

Math talk: Yes. In 15 + 25 + 12 the ones 5 and 5 make 10, leaving only 2 more to add.

Unit 3 · Math in My World: The Book Fair

1. 35 2. 24 + 35 = 59 3. 77 4. 79 5. Monday, by 6 books

Unit 3 Review

Part A: 1. 59 2. 88 3. 78 4. 68

Part B: 46 → 66 → 70 → 73, so 46 + 27 = 73

Part C: 64; 92

Part D: 60 + 8

Part E: 5. 67 6. 60

Part F: 7. 85 books

Challenge: e.g. 45 + 35 = 80, where 5 + 5 makes a new ten.

Unit 4 · Lesson 1 — Subtracting Tens and Ones

Try it together: 1. 33 2. 37

On your own: 1. 24 2. 44 3. 48 4. 22 5. 11 6. 72 7. 34 books

Math talk: If the top ones digit is smaller, there are not enough ones to take away, so a ten must be regrouped first.

Unit 4 · Lesson 2 — Counting Back and Counting Up

Try it together: 61 − 37 = 24

On your own: 45 → 50 → 70 → 72, so 72 − 45 = 27; 1. 25 2. 45 3. 18 4. 35

Math talk: When the two numbers are close together, counting up takes only a few small jumps.

Answer Key *continued*

Unit 4 · Lesson 3 — Regrouping a Ten

Try it together: 26; 45

On your own: 46; 36; 35;   1. 16   2. 34

Math talk: One ten and ten ones are the same amount, so trading changes how the number looks, not how much it is.

Unit 4 · Lesson 4 — Checking with Addition

Try it together: 1. 37, 37 + 27 = 64   2. 32, 32 + 18 = 50

On your own: 1. yes   2. no (it is 54)   3. yes   4. 44   5. 24

Math talk: Work the subtraction again, watching the regrouping step most closely.

Unit 4 · Lesson 5 — Word Problems within 100

Try it together: 1. 72 − 45 = 27 empty seats

On your own: 2. 45 stickers   3. 35 riders   4. 20 pages

Math talk: Two things happen to the apples — some are sold and some arrive — so two calculations are needed.

Unit 4 · Math in My World: The Class Trip

1. 37   2. 15   3. 3   4. 52   5. Yes, exactly — 52 people for 52 seats.

Unit 4 Review

Part A: 1. 44   2. 52   3. 26   4. 30

Part B: 34; 25

Part C: 46 → 50 → 70 → 73, so 73 − 46 = 27

Part D: Yes. 47 + 37 = 84.

Part E: 5. 44 beads   6. 44 points

Challenge: e.g. 63 − 37 = 26, check 26 + 37 = 63.

Unit 5 · Lesson 1 — Hundreds, Tens, and Ones

Try it together: 142; 307

On your own: 1. 4 h 7 t 2 o   2. 3 h 0 t 8 o   3. 6 h 9 t 0 o   4. 513   5. 809

Math talk: Without the zero the 4 would slide into the tens place and the number would mean 46, not 406.

Unit 5 · Lesson 2 — Reading and Writing Numbers

Try it together: 1. two hundred sixty-four   2. seven hundred thirty

On your own: 1. 481   2. 906   3. 150   4. three hundred seventy-five   5. six hundred eight

Math talk: Two hundred five. The tens place is empty, so no tens word is said, but the 0 must still be written.

Unit 5 · Lesson 3 — Expanded Form

Try it together: 500 + 70 + 3; 200 + 40 + 6

On your own: 1. 800 + 10 + 2   2. 300 + 50 + 9   3. 700 + 4   4. 362   5. 509   6. 940

Math talk: The chart shows the digits in columns; expanded form shows the values written out and added.

Answer Key *continued*

Unit 5 · Lesson 4 — Skip-Counting to 1,000

Try it together: 250, 300, 350, 400, 450; 615, 620, 625, 630, 635

On your own: hundreds 300, 400, 500, 600, 700; tens 740, 750, 760, 770, 780; fives 185, 190, 195, 200, 205; 1. 528 2. 296

Math talk: 590 + 10 is 600 — the tens fill a new hundred, so the hundreds digit grows and the tens digit returns to 0.

Unit 5 · Lesson 5 — Comparing to 1,000

Try it together: 1. > 2. < 3. = 4. >

On your own: 1. < 2. > 3. = 4. > 5. > 6. <; order 285, 528, 582

Math talk: Among whole numbers, yes — 100 has three digits and is greater than the two-digit 99.

Unit 5 · Math in My World: The Recycling Drive

1. Class 2B (264) 2. 200 + 60 + 4 3. two hundred forty 4. 240, 246, 264 5. 250

Unit 5 Review

Part A: 324; 1. 5 h 8 t 6 o

Part B: 400 + 10 + 7; 2. 639

Part C: 3. two hundred seventy-five 4. 804

Part D: 420, 430, 440, 450, 460; 700, 800, 900, 1,000

Part E: 5. < 6. > 7. = 8. <

Challenge: greatest 753, least 357, and 753 > 357.

Unit 6 · Lesson 1 — Ten and Hundred More or Less

Try it together: 1. 283 2. 581 3. 640 4. 639

On your own: 1. 175 2. 502 3. 584 4. 736 5. 307 6. 603 7. 545

Math talk: The tens and ones digits stay the same; only the hundreds digit changes.

Unit 6 · Lesson 2 — Adding Hundreds and Tens

Try it together: 1. 368 2. 588

On your own: 1. 367 2. 780 3. 738 4. 748 5. 570 6. 687 7. 400 + 70 + 6, and 476 + 200 = 676

Math talk: Each place can be added on its own, so three easy sums replace one hard one.

Unit 6 · Lesson 3 — Adding with Regrouping

Try it together: 392; 622

On your own: 613; 463; 1. 415 2. 641

Math talk: It is worth one ten, because ten ones were traded for it and moved into the tens column.

Unit 6 · Lesson 4 — Subtracting with Regrouping

Try it together: 246; 275

On your own: 276; 344; 1. 253 2. 677

Math talk: A zero has nothing to give, so you must open a unit from two places to the left before you can regroup.

Answer Key *continued*

Unit 6 · Lesson 5 — Explaining Why It Works

Try it together: 1. 33, 333 2. 63, 263

On your own: 1. 250 is 25 tens, and 25 tens + 25 tens is 50 tens, which is 500. 2. Yes. He added 2 to both numbers, so the difference stays the same; 462 − 200 = 262. 3. 306 (use 406 − 100)

Math talk: Subtraction measures the gap between two numbers. Moving both numbers by the same amount slides the gap along without changing its size.

Unit 6 · Math in My World: The Library Count

1. 248 + 376 = 624 2. 181 3. 295 4. 819 5. 669

Unit 6 Review

Part A: 1. 472 2. 715

Part B: 381; 642

Part C: 3. 546 4. 780

Part D: 274; 252

Part E: 5. Yes. Two was moved from the 47 to the 298, so the total is unchanged; both equal 345.

Part F: 6. 457 pencils

Challenge: e.g. 432 − 256 = 176, check 176 + 256 = 432.

Unit 7 · Lesson 1 — Coins and Their Values

Try it together: 15¢; 50¢

On your own: 1. 1 2. 5 3. 10 4. 25; 25¢; 37¢

Math talk: Counting by 25 is less familiar than counting by 10, so quarters need more care.

Unit 7 · Lesson 2 — Counting Mixed Coins

Try it together: 60¢; 21¢

On your own: 1. 40¢ 2. 32¢ 3. 56¢ 4. they are equal (both 30¢)

Math talk: Starting with the largest coin lets you skip-count most of the way, leaving only small steps at the end.

Unit 7 · Lesson 3 — Dollars and Cents

Try it together: 1. 63¢ 2. $3

On your own: 1. 100 2. 20 3. 10 4. 4; 5. 100¢ ($1) 6. exactly one dollar

Math talk: The two symbols mean the same kind of thing, so using both at once would say the amount twice.

Unit 7 · Lesson 4 — Same Amount, Different Coins

Try it together: 35¢ could also be 3 dimes and 1 nickel, or 7 nickels.

On your own: 1. quarter, dime, nickel 2. two quarters and a nickel 3. two quarters, one dime, one nickel, three pennies 4. e.g. two quarters, or five dimes

Math talk: A quarter and a nickel — only two coins.

Answer Key *continued*

Unit 7 · Lesson 5 — Money Word Problems

Try it together: 1. 30 − 28 = 2¢

On your own: 2. 75¢ 3. 25¢ 4. 10¢ 5. No. He has 63¢, which is 7¢ short.

Math talk: Count up from the price to the amount paid, using coins for each step.

Unit 7 · Math in My World: The School Shop

1. 65¢ 2. Yes, with 5¢ left 3. 25 + 15 = 40¢ 4. 25¢ 5. a pencil (25¢)

Unit 7 Review

Part A: 1. 5 2. 25 3. 10 4. 1

Part B: 46¢; 60¢

Part C: 5. 10 6. 42¢

Part D: 7. one quarter, two dimes (3 coins)

Part E: 8. 20¢ 9. 30¢

Challenge: e.g. two quarters and two dimes; seven dimes; or two quarters, one dime, and two nickels (six coins: two quarters, one dime, two nickels — that is five, so use two quarters and four nickels for exactly six).

Unit 8 · Lesson 1 — Time to Five Minutes

Try it together: 1. 3:10 2. 9:40

On your own: 1. 1:15 2. 5:50 3. 11:35 4. on the 9

Math talk: The hour hand moves slowly all through the hour, so by half past it sits halfway to the next number.

Unit 8 · Lesson 2 — Quarter Past and Quarter To

Try it together: quarter past 8 (8:15); quarter to 3 (2:45)

On your own: 1. b 2. c 3. a; hands at 5:15 and at 8:45

Math talk: A digital clock counts minutes past the hour, and 6:45 is still inside the 6 o'clock hour.

Unit 8 · Lesson 3 — A.M. and P.M.

Try it together: 1. a.m. 2. p.m. 3. p.m.

On your own: 1. a.m. 2. a.m. 3. p.m. 4. a.m. 5. answers vary

Math talk: Noon is written 12:00 p.m. It is confusing because the 12 does not fit the usual pattern of counting up from 1.

Unit 8 · Lesson 4 — Analog and Digital

Try it together: 6:10; 11:20

On your own: 1. 2:05 2. 8:50 3. 12:30; hands with the long hand on 8 and the short hand between 3 and 4

Math talk: Digital is faster to read exactly; the analog face shows how much of the hour has gone and how much is left.

Unit 8 · Lesson 5 — Ordering Times

Try it together: 1. 3:15 p.m. 2. 7:00 a.m.

On your own: 1. 6:15, 6:30, 6:45 a.m. 2. 9:30 a.m., 12:00 p.m., 4:05 p.m. 3. recess 4. any time from 2:25 to 2:55 p.m.

Math talk: 3:00 a.m. and 3:00 p.m. look the same on the clock, so the label decides which is earlier.

Answer Key *continued*

Unit 8 · Math in My World: My School Day

1. 9:15 a.m. 2. recess 3. long hand on 9, short hand between 10 and 11 4. p.m. 5. three o'clock in the afternoon

Unit 8 Review

Part A: 1. 4:20 2. 10:55

Part B: 3. 6:15 4. quarter to nine

Part C: 5. a.m. 6. p.m.

Part D: hands at 2:35; hands at 4:45

Part E: 7. 8:40 a.m., 11:55 a.m., 1:10 p.m.

Challenge: 4:15 p.m. Each full turn of the long hand is one hour.

Unit 9 · Lesson 1 — Measuring in Inches

Try it together: 1. 5 inches 2. 2 inches

On your own: 1. 6 inches 2. 4 inches 3. answers vary

Math talk: The measurement would be one inch too short, because the first inch was never counted.

Unit 9 · Lesson 2 — Measuring in Centimeters

Try it together: 1. 12 cm 2. 6 cm

On your own: 1. 9 cm 2. 14 cm 3. more — centimeters are smaller, so more of them are needed

Math talk: More. A smaller unit always gives a larger count for the same length.

Unit 9 · Lesson 3 — Choosing a Unit

Try it together: 1. inches (or centimeters) 2. feet 3. yards or meters

On your own: 1. centimeters 2. feet 3. meters or yards 4. inches 5. 12 6. 100

Math talk: The number would be enormous and hard to work with, even though it would not be wrong.

Unit 9 · Lesson 4 — Estimating Lengths

Try it together: 1. 7 in. 2. 4 m

On your own: 1. 9 in. 2. 5 m 3. 8 cm 4. 10 m 5. answers vary

Math talk: An estimate warns you if the measurement went wrong, and helps you pick the right tool and unit.

Unit 9 · Lesson 5 — Comparing Lengths

Try it together: 1. $40 - 25 = 15$ cm

On your own: 2. 1 cm 3. 61 in. 4. 15 m 5. 120 cm

Math talk: Add. Joining lengths end to end makes a total, not a difference.

Unit 9 · Math in My World: The Class Race

1. Ben and Dan (25 feet) 2. 13 feet 3. Ben and Dan 4. $18 + 12 = 30$ feet 5. Feet — inches would give very large numbers for distances this long.

Answer Key *continued*

Unit 9 Review

Part A: 1. 5 inches 2. 11 cm

Part B: 3. inches (or centimeters) 4. meters or yards

Part C: 5. 2 m 6. 6 in.

Part D: 7. 19 cm 8. 61 in.

Part E: 9. 12 10. 100

Challenge: the pencil is 11 cm, so the ruler is 22 cm.

Unit 10 · Lesson 1 — Line Plots

Try it together: 1. 4 cm 2. 13 ribbons

On your own: 1. 3 2. 3 3. 6 cm (only one) 4. 4 5. answers vary

Math talk: A tall stack means many objects share that measurement, so it is the most common length.

Unit 10 · Lesson 2 — Picture Graphs

Try it together: 1. 6 2. 4 3. 20

On your own: 1. 12 2. 6 3. 26 4. 8

Math talk: A larger key keeps the rows short when the numbers are big, so the whole graph still fits on the page.

Unit 10 · Lesson 3 — Bar Graphs

Try it together: 1. 6 2. blue 3. 26

On your own: 1. 8 2. 5 3. 16 4. 7

Math talk: If bars started at different places their heights could not be compared fairly.

Unit 10 · Lesson 4 — Comparing on Graphs

Try it together: 1. 2 2. 33

On your own: 1. 14 2. 8 3. 14 4. 34 5. bus and bike (10 + 4 = 14)

Math talk: Words such as *how many more*, *fewer*, and *difference* point to subtraction.

Unit 10 · Lesson 5 — Collecting Data

Try it together: 9; 13

On your own: answers vary; the total in question 3 should be 10.

Math talk: A tally is one quick stroke, and the groups of five make the final count fast.

Unit 10 · Math in My World: The Weather Chart

1. 11 2. 5 3. 11 4. 28 5. cloudy and windy (8 + 3 = 11)

Unit 10 Review

Part A: 1. 5 inches 2. 14

Part B: 3. 10 4. 4

Part C: 5. C 6. 5 7. 20

Part D: 12

Challenge: The other two add to 13, but 13 cannot be split into two equal whole numbers — so no whole-number answer exists. Accept any explanation that spots this.

Answer Key *continued*

Unit 11 · Lesson 1 — Sides and Angles

Try it together: 1. 3 sides, 3 angles 2. 6 sides, 6 angles

On your own: trapezoid; pentagon; rhombus. 1. hexagon 2. triangle 3. circle 4. a pentagon drawn

Math talk: Yes. A square has four sides with square corners, which is what a rectangle needs; a square is a special rectangle with all sides equal.

Unit 11 · Lesson 2 — Faces and Edges

Try it together: sphere; cone; pyramid

On your own: 1. 6 2. 12 3. 8 4. a cylinder 5. a cone 6. a sphere

Math talk: Spheres, cones, and cylinders roll. Each has a curved surface.

Unit 11 · Lesson 3 — Drawing Shapes

Try it together: a triangle; a hexagon

On your own: 1. a square (or rhombus) 2. a pentagon 3. a circle 4. answers vary, e.g. a door or a book

Math talk: Yes — a trapezoid or a rhombus has 4 sides but does not have four square corners.

Unit 11 · Lesson 4 — Rows and Columns of Squares

Try it together: 1. 4 rows, 3 columns, 12 squares

On your own: 1. 12 2. 9 3. 8 squares

Math talk: Yes — the total is 12 either way. Turning the rectangle a quarter turn swaps the rows and columns but not the total.

Unit 11 · Lesson 5 — Arrays and Repeated Addition

Try it together: 7 + 7 = 14

On your own: 1. 15 2. 16 3. 20 muffins 4. 7 + 7 + 7 = 21

Math talk: In repeated addition every addend is the same number, which is what makes an array quick to count.

Unit 11 · Math in My World: The Tile Floor

1. 6 2. 4 3. 6 + 6 + 6 + 6 = 24 4. 24 tiles 5. a square, with 4 angles

Unit 11 Review

Part A: hexagon; triangle; trapezoid

Part B: cube; cylinder; pyramid. 1. 6 faces, 12 edges

Part C: 2. a pentagon drawn

Part D: 3. 3 rows, 5 columns, 15 squares

Part E: 3 + 3 + 3 + 3 = 12

Challenge: 5 columns, because 5 + 5 + 5 + 5 = 20.

Unit 12 · Lesson 1 — Halves

Try it together: the circle and the two-part rectangle

On your own: 1. 2 2. whole 3. No. A half is always half of its own whole, so half a large pizza is bigger.

Math talk: No. Halves must be equal in size; unequal parts are just two pieces.

Answer Key *continued*

Unit 12 · Lesson 2 — Thirds

Try it together: thirds; halves; thirds

On your own: 1. 3 2. whole 3. 3 4. e.g. three horizontal strips instead of three vertical ones

Math talk: One half. Splitting a whole into fewer shares makes each share larger.

Unit 12 · Lesson 3 — Fourths

Try it together: fourths; fourths; halves

On your own: 1. 4 2. 2 3. one third 4. fourth (quarter)

Math talk: A half is one of two equal shares, and putting two fourths together covers exactly that much of the whole.

Unit 12 · Lesson 4 — Same Whole, Different Shapes

Try it together: yes — both show fourths of the same square

On your own: 2. yes 3. no 4. No. The wholes were different sizes, so half of the large cake is more.

Math talk: Yes. The sandwiches are identical, so each half is the same amount however it was cut.

Unit 12 · Lesson 5 — Equal Shares Every Day

Try it together: 1. half 2. fourth 3. third

On your own: 1. 2 fourths 2. yes 3. 2 thirds 4. a rectangle split into three equal strips

Math talk: If the shares are unequal, someone receives more than someone else, which is not fair sharing.

Unit 12 · Math in My World: The Birthday Party

1. fourth 2. third 3. half 4. the melon (a half is the largest share) 5. two fourths, or one half 6. one of the four parts shaded

Unit 12 Review

Part A: halves; thirds; fourths

Part B: one of three parts shaded; one of four parts shaded

Part C: 1. 4 2. 3

Part D: 3. one half 4. half

Part E: 5. Yes. The wholes are identical and each is split into two equal shares, so the halves are the same size even though they are different shapes.

Part F: 6. third

Challenge: e.g. four vertical strips, or two rows of two squares. Both give four equal shares of the same rectangle.

Standards Index

Find every lesson that teaches a given standard. Lesson 3.2 means Unit 3, Lesson 2.

Standards Index *continued*

About the Author

Emma Brightwell writes mathematics books for children in the elementary grades. Her work focuses on making the reasoning behind mathematics visible — not just the answer, but the picture and the thinking that produced it.

The New Generation Mathematics series covers every elementary and secondary grade. Each grade has a full-color textbook and a matching workbook, so children can learn an idea and then practice it without hunting for extra material.

Every book in the series is written to the United States Common Core State Standards for Mathematics, with the standard code printed on every lesson.

Certificate of Achievement

This certifies that

has completed Grade 2 Mathematics
with curiosity, effort, and careful thinking.

Date

Teacher or family

Well done, mathematician. Grade 3 is waiting.

www.ingramcontent.com/pod-product-compliance
Lightning Source LLC
LaVergne TN
LVHW010720061126
851857LV00014B/1273

* 9 7 8 0 5 5 7 9 2 1 0 2 7 *